The R.A.M.S. Library of Alchemy

Volume 34

Alchemy Rediscovered and Restored

by
Archibald Cockren

R.A.M.S. Publishing Company

Alchemy Rediscovered and Restored

by
Archibald Cockren

Produced by

Restorers of Alchemical Manuscripts Society

R.A.M.S. Publishing Company

R.A.M.S. Publishing Company
117 Rutherford Lane
Stuarts Draft VA 24477

Alchemy Rediscovered and Restored

First Edition 2015

ISBN-13 **978-1511904681**
ISBN-10 **1511904682**

Image Processing by Philip N. Wheeler

This book is sold for informational purposes only. Neither the publisher nor the editor shall be held accountable for the use or misuse of the information in this book.

Printed in the United States of America

Table of Contents

Dedicated to Hans W. Nintzel,

American Alchemist

and

Founder of the

Restorers of Alchemical Manuscripts Society

(R.A.M.S.)

Disclaimer

Liability: The publisher does not warrant or assume any legal liability or responsibility for the accuracy, completeness, or usefulness of any information, apparatus, product, or process disclosed. The publisher makes no representation as to the accuracy or completeness of the contents of this book and specifically disclaims any implied warranty of merchantability or fitness for a particular purpose. No warranty may be created or extended by written sales materials or sales representatives. You should obtain professional consultation where appropriate. The publisher shall not be liable for any loss of profit or other commercial or personal damages, including but not limited to special, incidental, consequential, or other damages.

Introduction

Philip N. Wheeler

Archibald Cockren was a practicing physician of the 20th century. Disenchanted with the medical treatments of his day, Dr. Cochren turned to the study of Alchemy. As shown in this book, he studied the works of many of the same authors frequently included in The R.A.M.S. Library of Alchemy. The history of these great Alchemists is very informative. This is a very readable 20th century text on Alchemy that was recommended by Hans W. Nintzel.

This volume also contains the short works:

- The Smaragdine Table of Hermes Trismegistus
- 'Tractatus Aureusi" or The Golden Tractate of Hermes
- Book of the Revelation of Hermes, interpreted by Theophrastus Paracelsus

ALCHEMY REDISCOVERED AND RESTORED

By Archibald Cockren

Philadelphia, David McKay

1941

With an account of the extraction of the seed of metals and the preparation of the medicinal elixir according to the practice of the hermetic Art and of the Alkahest of the Philosopher

Produced by

R.A.M.S.

Restorers of Alchemical Manuscripts

THE SMARAGDINE TABLE

OF HERMES TRISMEGISTUS

said to be found in the Valley of Ebron, after the Flood.

1. I speak not fiction, but what is certain and most true.

2. What is below is like that which is above, and what is above is like that which is below for performing the miracle of one thing.

3. And as all things were produced from One by the Mediation of One, so all things are produced from this One thing by adaptation.

4. Its father is the Sun, its mother was the Moon, the wind carried it in its belly, its nurse is the Earth.

5. It is the cause of all perfection throughout the whole world.

6. Its power is perfect if it be changed into the earth.

7. Separate the earth from the fire, the subtle from the gross, gently and with judgment.

8. It ascends from earth to heaven, and descends again to earth, thus you will possess the glory of the whole World and all obscurity will fly away.

9. This thing is the fortitude of all fortitude, because it overcomes all subtle things, and penetrates every solid thing.

10. Thus were all things created.

11. Thence proceed wonderful adaptations which are produced in this way.

12. Therefore am I called Hermes Trismegistus, possessing the three parts of the philosophy of the whole World.

13. What I had to say concerning the operation of the Sun is complete.

FOREWORD

BY SIR DUDLEY BORRON MYERS, O.B.E.

Having been intimately associated with Archibald
Cockren during the past ten years, and having long
since learnt to place implicit confidence in his
efficiency and reliability in all matters to which
he has devoted his many remarkable gifts and
talents, it affords me real pleasure to write a few
words by way of introduction to 'Alchemy
Rediscovered and Restored.'

In this book he tells of the sensational work which
he has accomplished in once more bringing to light,
and to the service of humanity, secrets which
baffled the majority of scientists of all ages, and
which, for several centuries, have been buried in a
grave of doubt and skeptical tradition. That this
grave should at last have been opened, and that the
real, albeit hidden secrets which it contained
should now stand revealed and proclaimed, must
undoubtedly be regarded as an epoch-making event.

I do not myself claim to have any scientific
knowledge whatever, but seeing is believing, and I
have been privileged to keep in close touch with the
author's experiments from the very beginning. Not
only have I seen the results achieved, but I, among
many others, have been able to test and pay grateful
tribute to the efficacy of the Elixirs produced by
the alchemical process. These, one may venture to
assert, cannot fail as they become better known to
prove a very valuable addition to the remedies at
present available to mankind.

There is no question of the claims which are put forward in this book being taken on trust. On the contrary they are open to the fullest examination. The proofs are there and they can safely be left to speak for themselves, in the light of the outcome of any investigations to which they may be subjected.

Seeing the far-reaching importance of the author's researches and discoveries it is necessary that some account should be given of his career, and of those qualifications in the wide field of physiology which entitle him to consideration in questions of the treatment of human ailments.

After the necessary period of training he was, in 1904, certificated at the National Hospital for Paralysis and Epilepsy as fully qualified for all purposes of massage, remedial exercises, and electrical treatment. From this hospital he passed on to the staff of the Great Northern Central Hospital, where he remained for several years. From 1908 onwards, however, he was able to devote part of his time to the private practice in which he then for the first time established himself in the West End of London. This practice had necessarily to be given up during the War.

The years 1915 and 1916 found him in complete charge of all electrical, massage, manipulative, and remedial exercises at the Russian Hospital for British Officers in South Audley Street, London. This hospital, it may be stated, was opened by the Russian nobility resident in London, and was wholly maintained by Russian money. From there he passed on in a similar capacity (1917--18) to the Prisoners of

War Hospital. He was at the same time attached to the Millbank Military Hospital. In 1918, he was transferred to the Australian Army, and was on the Peace Conference Staff of the Australian Prime Minister in 1919. Since then, that is to say for the past twenty years, he has been in permanent private practice in the West End of London.

For over twenty years he has been a keen student of the sciences of metallurgy, No-chemistry, and bacteriology, and it will thus be seen that in the claims he now advances in this book he writes with that measure of authority which a life devoted to the alleviation of suffering, and to the effective treatment of human ailments, undoubtedly confers on him.

It is given to few men to make such momentous discoveries as have rewarded his persistent work and patience. His work has, indeed, to my knowledge, often been pursued under conditions of great difficulty and disappointment. May what he has accomplished in the interests of science and of the human race bring him the reward which he deserves-- the reward of general recognition and appreciation of the results achieved.

DUDLEY B. MYERS

PART I: HISTORICAL

CHAPTER I: BEGINNINGS OF ALCHEMY

To most of us the word 'alchemy' calls up the picture of a medieval and slightly sinister laboratory in which an aged, black-robed wizard brooded over the crucibles and alembics that were to bring within his reach the Philosophers' Stone, and with that discovery the formula for the elixir of life and the transmutation of metals. But one can scarcely dismiss so lightly the science--or art, if you will--which won to its service the lifelong devotion of men of culture and attainment from every race and clime over a period of hundreds, or, indeed, thousands, of years, for the beginnings of alchemy are hidden in the mists of time. Such a science is something far more than an outlet for a few eccentric old men in their dotage.

What was the motive behind the constant strivings, the never-failing patience in the unravelling of the mysteries, the tenacity of purpose in the face of persecution and ridicule through the countless ages that led the alchemist to pursue undaunted his appointed way? Something far greater, surely, than a mere vainglorious desire to transmute the base metals into gold, or to brew a potion to prolong a little longer this earthly span, for the devotees of alchemy in the main cared little for these things. The accounts of their lives almost without exception lead us to believe that they were concerned with things spiritual rather than with things temporal. Rather were these men inspired by a vision, a vision of man made perfect, of man freed from disease and the limitations of warring faculties both mental and physical, standing as a god in the realization of a power that even at this very moment of time is lying hidden in the deeper strata of his consciousness, a vision of man made truly in the image and likeness of the one Divine Life in all its Perfection, Beauty, and Harmony.

To appreciate and understand these adepts visions it is necessary to trace to some extent the history of their cult, so let us for a space step back into the past to catch a glimpse of these men, of their work and ideals, and more important still, of the possibilities that their life-work might bring to those who to-day are seeking for fuller knowledge and wider horizons.

References are to be found in the myths and legends of China. From a book written by Edward Chalmers Werner, a late member of the Chinese Government's Historiological Bureau, Peking, comes this quotation from old Chinese records:

> Chang Tao-Ling, the first Taoist pope, was born in A.D. 35 in the reign of the Emperor Kuang Wu Ti of the Han dynasty. His birthplace is variously given as the T'ien-mu Shan, "Eye of Heaven Mountain," in Lin-an-Hsien in Chekiang, and Feng-yang Eu in Anhui. He devoted himself wholly to study and meditation, declining all offers to enter the service of the State. He preferred to take up his abode in the mountains of Western China where he persevered in the

study of alchemy and in cultivating the virtues of purity and mental abstraction. From the hands of Lao Tzu he received supernaturally a mystic treatise, by following the instructions in which he was successful in his search for the Elixir of Life.

This reference demonstrates that alchemy was studied in China as early as the commencement of the Christian era, so that its origin must probably lie far back in Chinese history.

From China we must now travel to Egypt, whence alchemy as known in the West seems to have sprung. The great Egyptian adept king, named by the Greeks *Hermes Trismegistus,*[1] is thought to have been the founder of the art. Reputed to have lived about 1900 B.C., he was highly celebrated for his wisdom and skill in the operation of nature, but of the works attributed to him only a few fragments escaped the destroying hand of the Emperor Diocletian in the third century A.D., namely, the Asclepian Dialogues and the Divine Poemanda. If we may judge from these fragments (both preserved in the Latin by Fianus and translated into English by Dr. Everard) it would seem to be of inestimable loss to the world that none of these works have survived in their entirety. The famous Smaragdine Table of Hermes (Tabula Smaragdina) I have placed at the beginning of this book, for although it would be difficult to prove its origin, yet it still represents a good example of Hermetic phraseology. There have been various stories of the origin of the Tract, one being that

---

[1] See The R.A.M.S. Library of Alchemy Volume 4, which includes The Golden Work of Hermes Trismegistus and Hermes Unveiled by Cyliani.

the original emerald slab upon which the precepts were said to be inscribed in Phoenician characters was discovered in the tomb of Hermes by Alexander the Great. In the Berne edition (1545) of the Summa Perfectionis the Latin version is printed under the heading:

'The Emerald Tables of Hermes the Thrice Great concerning Chymistry, Translator unknown. The words of the Secrets of Hermes which were written on the Tablet of Emerald found between his hands in a dark cave wherein his body was discovered buried.'

An Arabic version of the text was discovered in a work ascribed to Jabir, which was probably made about the ninth century. In any case it must be one of the oldest alchemical fragments known, and that it is a piece of Hermetic teaching I have no doubt, as it corresponds to teaching in the Poemanda and 'Fragments of a Faith Forgotten' in relation to the teaching of the thrice-greatest Hermes. It also teaches the unity of matter and the truth that all form is a manifestation from one root, the Aether, which teaching corroborates the theory of our present-day scientists. This table, in conjunction with the Tractatus Aureus or Golden Treatise which I have inserted at the end of this book, is well worth reading, particularly in the light of my elucidation of the general alchemical symbolism. Unhappily, it is all that remains to us of the Egyptian sacred art.

The third century A.D. seems to have been a period when the science was widely practiced, but it was also during this century, in the year 296, that Diocletian sought out and burnt all the Egyptian books on alchemy and the other occult sciences, and in so doing destroyed all evidence of progress made up to that date. In the fourth century *Zosimus the Panopolite* wrote his express treatise on 'The Divine Art of Making Gold and Silver,' and in the fifth *Morienus,* a hermit of Rome, left his native city and set out to seek the sage *Adfar,* a solitary adept whose fame had reached him from Alexandria. He found him, and after gaining his confidence became his disciple. After the death of his patron Morienus came into touch with King Calid, and a very attractive work purporting to be a dialogue between himself and the King is still extant under the name of Morienus. In this century *Cedrenus* also appeared, a magician who professed alchemy.

The next name of note, that of *Geber,*[2] occurs in or about A.D. 750. Geber's true name was Abou Moussah Djfar--Al Sofi, or The Wise. Born at Houran in Mesoptamia, he is generally esteemed by adepts as the greatest of them all after Hermes. Of the five hundred treatises said to have been composed by him only three remain to posterity--'The Sum of the Perfect Magistery,' 'The Investigation of Perfection,' and his' Testament.' It is to him, too, that we are indebted for the first mention of corrosive sublimate, red oxide of mercury and nitrate of silver. Skillfully indeed did Geber veil his discovery, for from his mysterious style of writing we derive the word' geber' or gibberish, but those who have really understood Geber, his adept compeers, declare with one accord that he has declared the truth, albeit disguisedly, with great acuteness and precision.

---

[2] See The R.A.M.S. Library of Alchemy Volume 9, Summa Perfectionis by Geber (Abu Mūsā Jābir ibn Hayyān).

Rhasis, another Arabian alchemist, became famous for his practical displays in the art of transmutation of base metals into gold.

In the tenth century *Al Farabi* enjoyed the reputation of being the most learned man of his age, and another great alchemist of this century was *Avicenna,* whose real name was Ebu Cinna. Born at Bokara in A.D. 980, he was the last of Egyptian Philosophers of note.

CHAPTER II: EARLY EUROPEAN ALCHEMISTS

About the period of the first Crusades alchemy shifted its centre to Spain, to which country it had been introduced by the Moors. In the twelfth century *Artephius*[3] wrote 'The Art of Prolonging Human Life,' and is reported to have lived throughout a period of one thousand years. He himself affirms this:

> I, Artephius, having learnt all the art in the book of Hermes, was once as others, envious, but having now lived one thousand years or thereabouts (which thousand years have already passed over me since my nativity, by the grace of God alone, and the use of this admirable quintessence), as I have seen, through this long space of time, that men have been unable to perfect the same magistery on account of the obscurity of the words of the philosophers, moved by pity and good conscience, I have resolved, in these my last days, to publish in all sincerity and truly, so that men may have nothing more to desire concerning this work. I except one thing only, which is not lawful that I should write, because it can be revealed truly only by God, or by a master. Nevertheless, this likewise may be learned from this book, provided one be not stiff-necked and have a little experience.

Of the thirteenth-century literature, a work called 'Tesero' was attributed to *Alphonso,* King of Castile in 1272: *William de Loris* wrote 'Le Roman de Rose' in about 1282, assisted by *Jean de Meung,* who also

---

[3] See The R.A.M.S. Library of Alchemy Volume 3: Artephius, His Secret Book.

wrote 'The Remonstrance of Nature to the Wandering Alchemist,' and 'The Reply of the Alchemist to Nature.' *Peter d'Apona,* born near Padua in 1250, wrote several books on 'magic,' and was accused by the Inquisition of possessing seven spirits, each enclosed in a crystal vessel, who taught him the seven liberal arts and sciences. He died upon the rack.

Among other famous names appearing about this period is that of *Arnold de Villeneuve* or Villanova, whose most famous work is found in the 'Theatrum Chemicum.' He studied medicine in Paris, but was also a theologian and alchemist. Like his friend, Peter d'Apona, he was thought to obtain his knowledge from the devil and was charged by many with magical practices. Although he did not himself fall into the hands of the Inquisition, his books were condemned to be burnt in Tarragona by that body on account of their heretical content. For Villanova maintained that works of faith and charity were more acceptable in the eyes of God than the Sacrificial Mass!

The authority of *Albertus Magnus* (1234--1314) is undoubtedly to be respected, since he renounced all material advantages to devote the greater part of a long life to the study of philosophy in the seclusion of a cloister. When Albertus died, his fame descended to his 'sainted pupil' *Aquinas,* who in his 'Thesaurus Alchimae' to his friend the Abbot Reginald, speaks openly of the successes of Albertus and himself in the art of transmutation.

Raymond Lully is one of the alchemists about whose life there is so much conflicting evidence that it is practically certain that his name was used as a cover by a second adept either at the same or a later period. He was probably born in Majorca about 1235, and after a somewhat dissolute youth, he was induced, apparently by the tragic termination of an unsuccessful love affair, to turn his thoughts to religion. He became imbued with a burning desire to spread the gospel among the followers of Mohammed, and to this end devoted years to the study of Mohammedan writings, the better to refute the Moslem teachings. He travelled widely, not only in Europe, but in Africa and Asia, where his religious zeal nearly cost him his life on more than one occasion. He is said to have become acquainted with Arnold de Villanova and the Universal Science somewhat late in life, when his study of alchemy and the discovery of the Philosophers' Stone increased his former fame as a zealous Christian.

According to one story his reputation eventually reached *John Cremer,* Abbot of Westminster at the time, who after working at alchemy for thirty years, had still failed to achieve his aim, the Philosophers' Stone. Cremer therefore sought out Lully in Italy, and having gained his confidence, persuaded him to come to England, where he introduced him to Edward II. Lully, being a great champion of Christendom, agreed to transmute base metals into gold on condition that Edward carried on the Crusades with the money. He was given a room in the Tower for his work, and it is estimated that he transmuted 50,000 pounds worth of gold. After a time, however, Edward became avaricious, and to compel Lully to carry on the work of transmutation made him prisoner, although with Cremer's aid he was able to escape from the Tower and return to the

Continent. Records state that he lived to be one hundred and fifty years of age and was eventually killed by the Saracens in Asia. At that age he is reputed to have been able to run and jump like a young man.

The enormous output of writings attributed to Lully (they total about 486 treatises on a variety of subjects ranging from grammar and rhetoric to medicine and theology) also seems to suggest that the name Lully was merely a pseudonym.
It was about this time that the science fell into grave disrepute, for the alchemist's claim to transmute metals offered great possibilities to any rogue with sufficient plausibility and lack of scruple to exploit the credulity or greed of his fellow-men, and there proved to be no lack either of charlatans or victims. Rich merchants and others greedy for gain were induced to entrust to the alleged alchemists gold, silver, and precious stones--which they lost--in the hope of getting them multiplied, and Acts of Parliament were passed in England and Pope's Bulls issued over Christendom to forbid the practice of alchemy on pain of death, although Pope John XXII is said to have practised the art himself and to have enriched the public treasury by this means.

In the fourteenth century lived the two *Isaacs Hollandus,*[4] father and son, Dutch adepts, who wrote 'De Triplici Ordinari Exiliris et Lapidis Theoria' and 'Mineralia Opera Sue de Lapide Philosophico.' The details of their operations on metals are the most explicit that have been given, and because of

---

[4] See The R.A.M.S. Library of Alchemy Volume 26, The Mineral Work, and Volume 27, The Vegetable Work.

this very lucidity have been discounted. John Read, for instance, Professor of Chemistry, in his 'Prelude to Chemistry, an Outline of Alchemy,' dismisses the writing of the Hollandus pair in a few words, possibly because their clarity of detail led him to suspect a blind. Alas, how blind sometimes are our experts themselves.

CHAPTER III: THE STORY OF NICHOLAS FLAMEL

In the whole history of alchemy surely one of the most interesting stories is that of Nicholas Flamel (1330-1 418), the most successful and most celebrated of France's adepts, and I am accordingly giving in his own words the account of the discovery which proved be the turning point in his life: 'I, Nicholas Flamel, Scrivener, living in Paris in the year of our Lord 1399 in the Notary Street, near St. James of the Boucherie, though I learned not much Latin, because of the poverty of my parents who, notwithstanding, were even by those who envy me most, accounted honest and good people: yet by the blessing of God I have not wanted an understanding of the books of the philosophers, but learned them and attained to a certain kind of knowledge, even of their hidden secrets. For which cause's sake, there shall not any moment of my life pass wherein, remembering this so vast good, I will not render thanks to this my good and gracious God. After the death of my parents, I Nicholas Flamel, got my living by the art of writing, ingrossing and the like, and in the course of time there fell into my hands a gilded book, very old and large, which cost me only two florins. It was not made of paper or parchment as other books are, but of admirable rinds, as it seemed to me, of young trees; the cover of it was brass, well bound, and graven all over with a strange sort of letters, which I took to be Greek characters, or some such like. This I know, that I could not read them; but as to the matter that was written within, it was engraven, as I suppose, with an iron pencil, or graven upon the said bark leaves; done admirably well, and in fair neat Latin letters, and curiously coloured.

'The book contained thrice seven leaves, so numbered at the top of each folio, every seventh leaf having painted images and figures instead of writing. On the first of these seven leaves there was depicted a virgin who was being swallowed by serpents; on the second a Cross upon which a serpent was crucified; on the last a wilderness watered by many fair fountains, out of which came a number of serpents, running here and there. On the first written leaf the following words were inscribed in great characters of gold "Abraham the Jew, Prince, Priest, Levite, Astrologer and Philosopher, unto the Jewish nation scattered through France by the wrath of God, wishing health in the name of the God of Israel." 'Thereafter followed great execrations and maledictions, with the word Maranatha repeated over and over, poured forth against anyone who should glance within, unless he were priest or scribe.

'The person who sold me this book must have known its value as much and as little as I who bought it. My suspicion is that it was either stolen from the miserable Jews or found hidden somewhere in the old place of their abode. On the second leaf the said Abraham consoled his people, praying them to avoid vices and idolatry more than all and await with patience the Messiah to come, who would vanquish all kings of the earth and thereafter reign, with those who were his own, in eternal glory. Without doubt this Abraham was a man of great understanding. On the third and rest of the written leaves he taught them the transmutation of metals in plain words, to help his captive nation in paying tribute to Roman

Emperors and for other objects which I shall not disclose. He painted the vessels on the margin, discovered the colours, with all the rest of the work, but concerning the Prime Agent he uttered no word, advising them only that he had figured and emblazoned it with great care in the fourth and fifth leaves. But all his skill notwithstanding, no one could interpret the designs unless he was far advanced in Jewish kabalah and well-studied in the book of the Philosophers. It follows that the fourth and fifth leaves were also without writing but full of illuminated figures exquisitely designed. On the obverse of the fourth leaf there was shewn a young man with winged feet having in his hand a caducean rod, encompassed by two serpents, and with this he stroke upon a helmet which covered his head. I took him to represent the Greek God Mercury. Unto him came running and flying with open wings a very old man, having an hour glass set upon his head and a scythe in his hands, like the figure of death, with which scythe he would have struck off the feet of Mercury. On the reverse of the fourth leaf a fair flower was depicted on the summit of a very high mountain, round which the North wind blustered. The plant had a blue stem, white and red flowers, leaves shining like fine gold, while about it the dragons and griffins of the North made their nests and their dwellings. On the obverse side of the fifth leaf there was a rose bush in flowers, in the midst of a fair garden, and growing hard by a hollow oak tree. At the foot bubbled forth a spring of very white water, which ran headlong into the depths below, passing first through the hands of a great concourse of people who were digging up the ground in search of it, save one person only, who paid attention to its weight. On the reverse side appeared a king carrying a great faulchion who caused his soldiers to destroy in his presence a multitude of little children, the mothers weeping at the feet of the murderers. The streams of blood were gathered by other soldiers into a great vessel, wherein the sun and moon bathe. Now, seeing that the history appeared to depict the slaughter of the innocents by Herod, and that I learned the main part of the Art in this book, it came about that I placed in their cemetery these hieroglyphic symbols of the Sacred Science.

'I have now described the content of the first five leaves, but I shall say nothing of all that was written in fair and intelligible Latin on the other pages, lest God should visit me for a greater wickedness than that of him who wished that all mankind had but one head so that he could cut it off at a blow. The precious book being in my possession I did little but study it night and day till I attained a fair understanding of all its processes, knowing nothing, however, respecting the matter of the work. I could therefore make no beginning and the result was that I became very sad and depressed. My wife Peronelle, whom I had married recently and loved as much as myself, was astonished and concerned greatly, endeavouring to comfort me and desiring earnestly to know whether she could not help me in my distress. I was never one who could hold his tongue and not only told her everything but showed her the book itself, for which she conceived the same affection as my own, taking great delight in the beautiful cover, the pictures and inscriptions, all of which she understood as little as I did. There was no small consolation, however, in talking with her about them and in wondering what

could be done to discover their meaning. At length I
caused the figures on the fourth and fifth leaves to
be painted as well as I could and had them put up in
my workroom, where I shewed them to many scholars in
Paris; but these also could throw no light upon
them. I went so far as to tell them that they had
been found in a book about the Philosophers' Stone,
but most of them made a mock of it and also of me.
An exception however was one named Anselm, a
licentiate of medicine and a deep student of the
Art. He desired earnestly to see my book and would
have done anything to have his way in the matter,
but I persisted in saying that it was not in my
possession, though I gave him a full account of the
process described therein.

'He declared that the first figures represented
time, which devours all things, while the six
written leaves shewed that a space of six years was
required to perfect the Stone, after which there
must be no further coction. When I pointed out that
according to the book the figures were designed to
teach the First Matter he answered that the six
years coction was like a second agent; that as
regards the first it was certainly shewn forth as a
white and heavy water, which was doubtless
quicksilver. The feet of this substance could not be
cut off, meaning that it could not be fixed and so
deprived of volatility except by such long decoction
in the pure blood of young children. The quicksilver
uniting with gold and silver in this blood would
change with them, firstly into a herb like that of
the fair flower on the reverse of the fourth leaf,
secondly by corruption into serpents, which
serpents, being dried and digested by fire, would

become Powder of Gold, and of such in truth is the Stone.

'This explanation sent me astray through a labyrinth of innumerable false processes for a period of one and twenty years, it being always understood that I made no experiments with the blood of children, for that I accounted villainous. Moreover, I found in my book that what the philosophers called blood is the mineral spirit in metals, more especially in gold, silver and quicksilver to the admixture of which I tended always. The licentiate's interpretation being more subtle than true, my processes never exhibited the proper signs at the times given in the book, so I was ever to begin again. At last, however, having lost all hope of understanding the figures, I made a vow to God and St. James that I would seek their key of some Jewish priest belonging to one of the Spanish synagogues. Thereupon, with the consent of Peronelle and carrying a copy of the figures, I assumed a pilgrim's weeds and staff, in the same manner as you see me depicted outside the said arch in the said churchyard where I put up the hieroglyphic figures, as also a procession representing on both sides of the wall and successive colours of the Stone which arise and pass off in the work, and the following inscription in French: "A procession is pleasing to God when it is done in devotion." These are the first words, or their equivalent, of a tract on the colours of the Stone by the King Hercules, entitled Iris, which opens thus "Operis Processio Multum Naturae Placet." I quote them for the benefit of scholars, who will understand the allusion. Having donned my pilgrim's weeds, I began to fare on the road, reaching Mountjoy and finally my destination at St. James,

where I fulfilled my vow with great devotion. On the
return journey I met with a merchant of Boulogne in
Leon, and to him I was indebted for acquaintance
with Master Candies, a doctor of great learning who
was Jewish by nation but now a Christian. When I
shewed him my copy of the figures he was ravished
with wonder and joy, and asked with great
earnestness whether I could give him news of the
book from which they were taken. He spoke in Latin
and I answered in the same language that if anyone
could decipher the enigma there was good hope of
learning its whereabouts. He began at once to
decipher the beginning.

'To shorten this part of the story he had heard much
talk of the work but as of a thing that was utterly
lost. I resumed my journey in his company,
proceeding from Leon to Ovideo and thence to
Sareson, at which port we set sail for France and
arrived in due time, after a prosperous voyage. On
our way to Paris my companion most truly interpreted
the major or part of my figures, in which he found
great mysteries, even to the points and pricks. But
unhappily when we reached Orleans this learned man
fell sick and was afflicted with extreme vomitings,
a recurrence of those from which he had suffered at
sea. He was continually in fear of my leaving him,
and though I was ever at his side he would still be
calling me. To my great sorrow he died on the
seventh day, and to the best of my ability I saw
that he was buried in the Church of Holy Cross at
Orleans. There he still lies, and may God keep his
soul, seeing that he made a good Christian end.

'He who would see the manner of my arrival home and
the satisfaction of Peronelle may look on us both as

we are painted on the door of the Chapel of St. James of the Boucherie hard by my house. We are shewn on our knees, myself at the feet of St. James of Spain and she at those of St. John, to whom she prayed so often. By the grace of God and the intercession of the Holy and Blessed Virgin, as also of the Saints just mentioned, I had gained that which I desired, being a knowledge of the First Matter, but not as yet of its initial preparation, a thing of all else most difficult in the world. In the end, however, I attained this also, after errors innumerable through the space of some three years, during which I did nothing but study and work as you will see me depicted outside the arch at the Chapel of St. James and St. John, ever praying to God rosary in hand, engrossed in a book, pondering the words of the philosophers and proving various operations suggested by their study. The fact of my success was revealed to me by the strong odour, and thereafter I accomplished the mastery with ease indeed I could scarcely miss the work had I wished, given a knowledge of the prime agents, their preparation and following my book to the letter. On the first occasion projection was made upon Mercury, of which I transmuted a half pound or thereabouts into pure silver, better than that of the mine, as I and others proved by assaying several times. This was done on a certain Monday, the seventeenth day of January 1392, Peronelle only being present. Thereafter, still following--word for word--the directions of my book, about five o'clock in the evening of the twenty-fifth day of the following April I made projection of the Red stone on the same amount of Mercury, still at my own house, Peronelle and no other with me, and it was duly transmuted into the same quantity of pure gold, much better than that of the ordinary metal, softer and more pliable. I speak in all truth. I have made it three times, with the aid of Peronelle, for she helped me in all my operations and understood the subject as well as myself. She could have done it alone without doubt, had she desired, and would have brought it to the same term. The first occasion gave me all that I needed, but I took great delight in contemplating the wonderful works of Nature within the vessels, and to signify that I made three transmutations you have only to look at the arch and the three furnaces depicted thereupon, answering to those which served in our operations.

'For a considerable time I was in no little anxiety lest Peronelle should prove unable to conceal her happiness and should let fall some words among her kinsfolk concerning our great treasure. I judged of her joy by my own, and great joy, like great sorrow is apt to diminish caution. But the most high God in His Goodness had not only granted me the blessing of the Stone, He had given me a chaste and prudent wife, herself endowed with reason, qualified to act reasonably, and more discreet and secret than other women are for the most part. Above all she was very devout and having no expectations of children, for we were now advanced in years, she began--like myself-- to think of God and to occupy herself with works of mercy. Before I wrote this commentary, which was towards the end of the year 1413, after the passing of my faithful companion, whom I shall lament all the days of my life, she and I had already founded and endowed fourteen hospitals, had built three Chapels and provided seven Churches with substantial gifts and revenues, as well as restoring their cemeteries.'

Nicholas Flamel died eventually in 1415 at the age of one hundred and sixteen years. Some evidence of his house, dating from 1407, is still to be seen in the building of 51, rue de Montmorency in Paris, and in the Musée de Cluny there is an inscribed tablet from his tomb in the old Church of St. Jaques-la-Boucherie, now demolished. This tablet, which is quite unique, had an interesting and somewhat chequered career. Lost for many years, after the demolition of St. Jacques-laBoucherie in 1717, it was eventually found in a shop in the rue des Arias, where the owner, a greengrocer and herbalist, had been using the smooth marble back as a chopping block for his herbs.

The tablet itself measures 58 x 45 centimetres, and is four centimetres thick. At the top is a carved representation of Christ, St. Peter, and St. Paul, and the inscription records that Nicholas Flamel, formerly a scrivener, left certain moneys and properties for religious and charitable purposes, including gifts to churches and hospitals in Paris.

I have retailed this account of Flamel's experiences in full as it seems to me to be of no mean interest, despite the fact that certain authorities have doubted its veracity. My own feeling about it is that the history is a true one; that the book of Abraham the Jew to which Flamel refers is evidently an allegorical writing of the whole process, and that the corresponding pictures are, to anyone versed in alchemical language, representative of the different phases of the work. Some writers and critics, certainly, have held these allegories up to ridicule as the outpourings of religious

visionaries, but here I think they demonstrate their
ignorance of the whole process. One of the greatest
proofs of the truth of this history is, in my
opinion, the point at which Flamel refers to the
attainment of the First Matter. Of this he says 'The
fact of my success was revealed to me by the strong
odour,' and this fact I myself have demonstrated in
the laboratory; the odour is unmistakable, and the
gas of such a volatile nature that it pervades the
whole house. In the theoretical and practical
sections I shall refer to this more fully.

CHAPTER IV: BASIL VALENTINE

RECORDS of the life of Basilius Valentinus, the Benedictine monk who for his achievements in the chemical sphere has been given the title of Father of Modern Chemistry, are a mass of conflicting evidence. Many and varied are the accounts of his life, and historians seem quite unable to agree as to his exact identity, or even as to the century in which he lived. It is generally believed, however, that 1394 was the year of his birth, and that he did actually join the Benedictine Brotherhood, eventually becoming Canon of the Priory of St. Peter at Erfurt, near Strasburg, although even these facts cannot be proved.

Whatever his identity, Basil Valentine[5] was undoubtedly a great chemist, and the originator of many chemical preparations of the first importance. Amongst these are the preparation of spirit of salt, or hydrochloric acid from marine salt and oil of vitriol (sulphuric acid) the extraction of copper from its pyrites (sulphur) by transforming it firstly into copper sulphate, and then plunging a bar of iron in the watery dissolution of this product: the method of producing sulpho-ether by the distillation of a mixture of spirit of wine and oil of vitriol: the method of obtaining brandy by the distillation of wine and beer, rectifying the distillation on carbonate of potassium.

In his writings he has placed on record many valuable facts, and whether Basil Valentine is the

---

[5] The R.A.M.S. Library of Alchemy Volume 1, Twelve Keys of Basilius Valentinus, and Of Natural & Supernatural Things.

correct name of the author or an assumed one matters little, since it detracts nothing from the value of his works, or the calibre of his practical experiments. From his writings one gathers that he was indeed a monk, and also the possessor of a mind and understanding superior to that of the average thinker of his day. The ultimate intent and aim of his studies was undoubtedly to prove that perfect health in the human body *is* attainable, and that the perfection of all metallic substance is also possible. He believed that the physician should regard his calling in the nature of a sacred trust, and was appalled by the ignorance of the medical faculty of the day whose members pursued their appointed way in smug complacency, showing little concern for the fate of their patients once they had prescribed their pet panacea.

The following quotation from Basil Valentine's 'Triumphal Chariot of Antimony'[6] is from the Latin version published at Amsterdam in 1685, and translated into English and published by James Elliott & Co., Falcon Court, Fleet Street, E.C., in 1893.

'. . . this quality of doctor,' he writes, 'cannot prepare his own medicines (such as they are) but must leave that work to another. He does not even know the colour of the remedies which he prescribes. He has not the slightest idea whether they are white or black, red or grey, blue or yellow, or whether the medicament is hot, cold, dry, or humid. He only knows one thing--that he has found the name of that medicine in his books, and pluming himself on the

---

[6] The R.A.M.S. Library of Alchemy Volume 2.

antiquity of his hoary knowledge, he claims the right of prior possession.

'Here again I am tempted to cry woe upon these foolish doctors whose consciences are seared with a hot iron, who do not care in the least for their patients, and will be called to a terrible account for their criminal folly on the day of judgment. Then they will behold Him whom they have pierced by neglecting their neighbour's welfare, while pocketing his money, and will see at last that they ought to have laboured night and day, in order to acquire greater skill in the healing of disease. Instead of this they complacently go on trusting to chance, prescribing the first medicine they happen to find in their books, and leaving the patient and the disease to fight it out as best they can. They do not even trouble to enquire in what way the medicines they prescribe are prepared. Their laboratory, their furnace, their drugs are at the Apothecary's, to whom they rarely or never go. They inscribe upon a sheet of paper, under the magic word "Recipe," the names of certain medicines, whereupon the Apothecary's assistant takes his mortar and pounds out of the wretched patient whatever health may still be left in him.

'Change these evil times, oh. God! Cut down these trees, lest they grow up to the sky! Overthrow these overweening giants, lest they pile mountain upon mountain and attempt to storm heaven! Protect the conscientious few who quietly strive to discover the mysteries of Thy creation! 'I will ask all my brothers in our Monastery to unite with me in earnest prayer, by day and by night, that God may enlighten the ignorance of these pseudo-doctors, that they may understand the virtues which he has implanted in created things, and may learn also that they can become manifest and operative only by means of that preparation which removes all harmful and poisonous impurities. I trust that God will answer our prayer, and that some of my brothers at least will survive to witness the blessed change which shall then take place on earth, when the thick veil of ignorance shall have been removed from the eyes of our opponents, and their minds shall have been enlightened to find the lost piece of silver. May God, who overrules the destinies of men, in His goodness and mercy bring about this consummation.'

On the subject of the perfection of metallic bodies, as in his reference to the Spagyric Art, the Grand Magistrum, the Universal Medicine, the Tinctures to transmute metals and other mysteries of the alchemist's art, he has completely mystified not only the lay reader, but the learned chemists of his own and later times. In all his works the important key to a laboratory process is apparently omitted. Actually, however, such a key is invariably to be found in some other part of the writings, probably in the midst of one of the mysterious theological discourses which he was wont to insert among his practical instructions, so that it is only by intensive study that the mystery can be unravelled. His most famous work is his 'Currus Triumphalis Antimonii' ('The Triumphal Chariot of Antimony'). It has been translated into German, French, and English, and has done more to establish his reputation as a chemist than any other. The best edition is undoubtedly that published at Amsterdam in 1671 with a commentary by Theodorus Kerckringius. In his preface Kerckringius states that he had actually spoken with Valentine besides studying his works. He speaks of Basil as 'the prince of all chemists, and the most learned, upright, and lucid of all alchemistic writers. He tells the careful student everything that can be known in alchemy; of this I can most positively assure you.' A perusal of this book makes it quite evident that Valentine had investigated very thoroughly the properties of antimony, and the findings on his experimental work with this metal have been brought forward as recent discoveries by chemists of our day.

His other works are 'The Medicine of Metals,' 'Of Things Natural and Supernatural,' 'Of the First Tincture, Root and Spirit of Metals,' 'The Twelve Keys,' and his 'Last Will and Testament.' It is alleged that this last work remained concealed for a number of years within the High Altar of the church belonging to the Priory. Such a story is quite feasible, since alchemists both before and after this era, deeming their works unfit for the age in which they were written, are known to have buried or otherwise secreted their writings for the discovery and benefit, as they doubtless hoped, of a more deserving and more enlightened age. Such manuscripts would very often not be discovered for several generations after the death of the author.

In view of his other outstanding achievements as a chemist of great ability, it seems not illogical to suppose that Valentine's Universal Method of Medicine should be capable of achieving as great a measure of success as his other somewhat more prosaic discoveries.

CHAPTER V: PARACELSUS

Philippus Aureolus Theophrastus Bombastus von Hohenheim, immortalized as Paracelsus,[7] was born in 1493. He was the son of a physician of repute, who has been described as a Grand Master of the Teutonic Order, and it was from him that Paracelsus took his first instruction.

At the age of sixteen he entered the University at Basle, where he applied himself to the study of alchemy, surgery, and medicine. With the science of alchemy he was already acquainted, having previously studied the works of Isaac Hollandus, whose writings roused in him the ambition to cure disease by medicine superior to the material at that time in use, for apart from his incursions into alchemy, Paracelsus is credited with the introduction of opium and mercury into medicine, while his works indicate an advanced knowledge of the science and principles of magnetism. These are some of the achievements which would seem to justify Manly Hall's description of him as 'the precursor of chemical pharmacology and therapeutics and the most original medical thinker of the sixteenth century.'

The Abbot Trithemius, an adept of a high order, and the instructor of the illustrious Henry Cornelius Agrippa, was responsible for Paracelsus' initiation into the science of alchemy. In 1516 he was still pursuing his research in mineralogy, medicine, surgery, and chemistry under the guidance of Sigismund Fugger, a wealthy physician of the city, but was forced to leave Basle hurriedly after

---

[7] See The R.A.M.S. Library of Alchemy Volume 6: Four Works of Paracelsus.

trouble with the authorities over his studies in necromancy. He started out on a nomad's life, supporting himself by astrological predictions and occult practices of various kinds.

His wanderings took him through Germany, France, Hungary, the Netherlands, Denmark, Sweden, and Russia. In Russia he is reported to have been taken prisoner by the Tartars and brought before the Grand Cham at whose court he became a great favourite. Finally, assuming this story to be true, he accompanied the Cham's son on an embassy from China to Constantinople, the city in which the supreme secret, the universal dissolvent, the alkahest, was imparted to him by an Arabian adept. For Paracelsus, as Manly Hall has said, gained his knowledge 'not from coated pedagogues, but from dervishes in Constantinople, witches, gipsies, and sorcerers, who invoked spirits and captured the rays of the celestial bodies in dew; of whom it is said that he cured the incurable, gave sight to the blind, cleansed the leper, and even raised the dead, and whose memory could turn aside the plague.' Paracelsus ultimately returned to Europe, passing along the Danube into Italy where he became an army surgeon. It was here apparently that his wonderful cures began. In 1526, at the age of thirty-two, he re-entered Germany, and at the university he had entered as a youth took a professorship of physics, medicine, and surgery. This was a position of some considerable importance, and was offered to him at the instance of Erasmus and Ecolampidus. Perhaps it was his behaviour at this time that eventually led to his title 'the Luther of physicians,' for in his lectures he made so bold as to denounce as antiquated the systems of Galen and his school,

whose teachings were held to be so unalterable and
inviolable by the authorities of that time, that the
slightest deviation from their teachings was
regarded as nothing short of heretical. As a
crowning insult he actually burnt the works of these
masters in a brass pan with sulphur and nitre! This
high-handed behaviour, coupled with his original
ideas, made him countless enemies. The fact that the
cures he performed with his mineral medicines
justified his teachings merely served further to
antagonize the medical faculty, infuriated at their
authority and prestige being undermined by the
teachings of a 'heretic' and 'usurper.' Thus
Paracelsus did not long retain his professorship at
Basle, but was forced once again to leave the city
and betake himself to a wanderer's life.

During the course of his second exile we hear of him
in 1526 at Colmar, and in 1530 at Nuremburg, once
again in conflict with the doctors of medicine, who
denounced him as an impostor, although once again he
turned the tables on his opponents by his successful
treatment of several bad cases of elephantiasis,
which he followed up during the next ten years by a
series of cures which were amazing at the period.

Franz Hartmann in his 'Paracelsus' says:
'He proceeded to Maehren, Kaernthen, Krain, and
Hungary, and finally to Salzburg, to which place he
was invited by the Prince Palatine, Duke Ernst of
Bavaria, who was a great lover of the secret art.
But he was not destined to enjoy a long time the
rest he so richly deserved. . .'

He died in 1541 after a short sickness in a small
room at the White Horse Inn near the quay, and his

body was buried in the graveyard of St. Sebastian. One writer supposes the event to have been accelerated by a scuffle with assassins in the pay of the orthodox medical faculty, but there is no actual foundation for this story.

Not one of his biographers seems to have found anything remarkable in the fact that at sixteen years of age Paracelsus was already well acquainted with alchemical literature. Even allowing for the earlier maturity of a man in those times, he must still have been something of a phenomenon in mental development. Certain it is that few of his contemporaries either could or would grasp his teachings, and his consequent irritation and arrogance in the face of their stupidity and obstinacy is scarcely to be wondered at. Although he numbered so many enemies among his fellow physicians, he also had his disciples, and for these no praise was too high for him. He was worshipped as their Noble and Beloved Monarch, the German Hermes, the Philosopher Trismegistus, Dear Preceptor and King, Theophrastus of Blessed Memory and Immortal Fame.

I am indebted to Mr. Arthur Edward Waite's translation from the German of the Hermetic and Alchemic Writings of Paracelsus for many of these facts of I life.

CHAPTER VI

ALCHEMY IN THE SIXTEENTH AND SEVENTEENTH CENTURIES

The first man to teach the chemistry of the human body and to declare, as did Paracelsus, that the true purpose of chemistry was the preparation of medicine for the treatment of disease was one *Jean Baptista van Helmont,* a disciple of Paracelsus, sometimes called the Descartes of Medicine.
In his treatise, 'De Natura Vitae Eternae,' he writes

'I have seen and I have touched the Philosophers' Stone more than once. The colour of it was like saffron in powder, but heavy and shining like pounded glass. I had once given me the fourth of a grain--I call a grain that which takes 600 to make an ounce. I made projection with this fourth part of a grain wrapped in paper upon eight ounces of quicksilver heated in a crucible. The result of the projection was eight ounces, lacking eleven grains, of the most pure gold.'

In his early thirties van Helmont retired to an old castle in Belgium near Brussels and remained there, almost unknown to his neighbours until his death in his sixty-seventh year. He never professed to have actually prepared the Philosophers' Stone, but gained his knowledge from alchemists he contacted during his years of research.

Van Helmont also gives particulars of an Irish gentleman named *Butler,* a prisoner in the Castle of Vilvord in Flanders, who during his captivity performed strange cures by means of the Hermetic medicine. The news of his cure of a Breton monk, a fellow-prisoner suffering from severe erysipelas, by the administration of almond milk in which he had merely dipped the Philosophers' Stone brought van Helmont, accompanied by several noblemen, post-haste to the Castle to investigate the case. In their presence Butler cured an aged woman of 'megrim' by dipping the Stone into olive-oil and then anointing her head. There was also an abbess who had suffered for eighteen years with paralysed fingers and a swollen arm. These disabilities were removed by applying the Stone a few times to her tongue.

In 'Lives of the Alchemystical Philosophers,' published in 1815, it is stated that prior to the events at Vilvord, Butler attracted some attention by his transmutations in London during the reign of James I. He is said to have gained his knowledge in Arabia and in this way. When a ship in which he had once taken passage was captured by African pirates, Butler was taken prisoner and sold into slavery in Arabia. His Arab master was an alchemical worker with knowledge of the correct processes. Butler assisted him in some of his operations, and when later he was able to make his escape from captivity, he carried off a large portion of the Red Powder.

Denys Zachare in his memoirs gives an interesting account of his pursuit of the Philosophers' Stone. At the age of twenty he set out to Bordeaux to undertake a college curriculum, and hence to Toulouse for a course of law. In this town he made the acquaintance of some students in possession of a number of alchemical books. It seems that at this time there was a craze for alchemical experiments among the students of Paris and other French towns, and this craze caught Zachare's imagination. His law studies were forsaken and his experiments in alchemy began. On his parents' death, having expended all his money on this new love of his he returned home and from their estate raised further money to continue his research. For ten years, according to his own statement, after experiments of all sorts and meetings with countless men with a method to sell, he sat down to study carefully the writings of the philosophers on the subject, and states that it was Raymond Lully's 'Testament, Codicil, and Epistle' addressed to King Robert that gave him the key to the secret. From the study of this book and 'The Grand Rosary' of Arnold de Villeneuve, he formulated a plan entirely different from any he had previously followed. After another fifteen months of toil he says:

'I beheld with transport the evolution of the three successive colours which testify to the True Work. It came finally at Eastertide; I made a projection of my divine Powder on quicksilver, and in less than an hour it was converted into fine gold. God knows how joyful I was, how I thanked him for this great grace and favour, and prayed for His Holy Spirit to pour yet more light upon mc that I might use what I had attained only to His praise and honour.'

In his one writing entitled 'Opusculum Chemicum' he gives his own personal narrative and states that the Art is the gift of God alone. The methods and possibilities of the transmutation of metals and the Tincture as a Medicine are also considered.

There is also the evidence of *John Frederick Helvetius,* as testified in 1666. He made claim to be an adept, but received the powder of transmutation from another. He writes:

'On December 27th, 1666, and in the forenoon, there came a certain man to my house who was unto me a complete stranger, but of an honest, grave and authoritative man, clothed in a simple garb like that of a Mennonite. He was of middle height, his face was long and slightly pock-marked, his hair was black and straight, his chin close-shaven, his age about forty-three or forty-four, and his native place North Holland, so far as I could make out. After we had exchanged salutations, he inquired whether he might have some conversation with me. It was his idea to speak of the Pyrotechnic Art, as he had read one of my tracts, being that directed against the Sympathetic Powder of Sir Kenelm Digby, in which I implied a suspicion whether the Great Arcanum of the Sages was not after all a gigantic hoax. He took therefore this opportunity of asking if indeed I could not believe that such a Grand Mystery might exist in the nature of things, being that by which a physician could restore any patient whose vitals were not irreparably destroyed. My answer allowed that such a Medicine would be a most desirable acquisition for any doctor and that none might tell how many secrets there may be hidden in Nature, but that as for me--though I had read much on the truth of this Art--it had never been my fortune to meet with a Master of Alchemical Science. I inquired further whether he was himself a medical man since he spoke so learnedly about the Universal Medicine, but he disclaimed my suggestion modestly, describing himself as a brass-founder, who had

always taken great interest in the extraction of
medicines from metals by means of fire. After some
further talk the Artist Elias--for he it was--
addressed me thus:

'"Seeing that you have read so much in the writings
of the alchemists concerning the Stone, its
substance, colour and wonderful effects, may I be
allowed to question whether you have yourself
prepared it."

'On my answering him in the negative he took from
his bag an ivory box of cunning workmanship in which
there were three large pieces of a substance
resembling glass or pale sulphur and informed me
that here was enough of the Tincture to produce
twenty tons of gold.

'When I held the treasure in my hands for some
fifteen minutes listening to an account of its
curative properties, I was compelled to return it,
not without a certain degree of reluctance. After
thanking him for his kindness I asked why it was
that his Tincture did not display that ruby colour
which I had been taught to regard as characteristic
of the Philosophers' Stone. He replied that the
colour made no difference and that the substance was
sufficiently mature for all practical purposes. He
refused somewhat brusquely my request for a piece of
his substance, were it no larger than a coriander
seed, adding in a milder tone that he could not do
so for all the wealth which I possessed; not indeed
on account of its preciousness but for another
reason that it was not lawful to divulge. Indeed, if
fire could be destroyed by fire he would cast it
rather into the flames. Then after a little

consideration he asked whether I could not shew him into a room at the back of the house, where we should be less liable to observation. Having led him into the state parlour, he requested me to produce a gold coin, and while I was finding it he took from his breast pocket a green silk handkerchief wrapped about five medals, the gold of which was infinitely superior to that of my own money. Being filled with admiration, I asked my visitor how he had attained this most wonderful knowledge in the world, to which he replied that it was a gift bestowed upon him freely by a friend who had stayed a few days at his house, who had taught him also how to change common flints and crystals into stones more precious than rubies, chrysolites and sapphires.

'"He made known to me further," said the artist, "the preparation of crocus of iron, an infallible cure for dysentry; of a metallic liquor, which was an efficacious remedy for dropsy, and of other medicines."

'To this, however, I paid no great heed as I, Helvetius, was impatient to hear about the Great Secret of all. The artist said further that his master caused him to bring a glass full of warm water to which he added a little white powder and then an ounce of silver, which melted like ice therein.

'"Of this he emptied one half and gave the rest to me. Its taste resembled that of fresh milk, and the effect was most exhilarating."

'I asked my visitor whether the potion was a preparation of the Philosophers' Stone, but he

replied that I must not be curious. He added
presently that at the bidding of his master he took
down a piece of lead water-pipe and melted it in a
pot, when the master removed some sulphurous powder
on the point of a knife from a little box, cast it
into the molten lead, and after exposing the
compound for a short time to a fierce fire he poured
forth a great mass of liquid gold upon the brick
floor of the kitchen.

'"The Master bade me take one-sixteenth of this gold
as a keepsake for myself and distribute the rest
among the poor, which I did by making over a large
sum in trust for the Church of Sparrendaur. In fine,
before bidding me farewell, my friend taught me this
Divine Art."

'When my strange visitor had concluded his
narrative, I besought him in proof of his statement
to perform a transmutation in my presence. He
answered that he could not do so on that occasion
but that he would return in three weeks and if then
at liberty to do so he would shew me something that
would make me open my eyes. He returned punctually
on the promised day and invited me to a walk, in the
course of which we spoke profoundly on the secrets
of Nature in fire, though I noticed that my
companion was exceedingly reserved on the subject of
the Great Secret. When I prayed him, however, to
entrust me with a morsel of his precious Stone, were
it no larger than a rape seed he delivered it like a
princely donation. When I expressed a doubt whether
it would be sufficient to tinge more than four
grains of lead he eagerly demanded it back. I
complied, hoping that he would exchange it for a
larger fragment, instead of which he divided it with

his thumb, threw half in the fire and returned the rest, saying, '"It is yet sufficient for you."'

The narrative goes on to state that on the morrow Helvetius prepared six drachms of lead, melted it in a crucible, and cast on the Tincture. There was a hissing sound and a slight effervescence, and after fifteen minutes Helvetius found that the lead had been transformed into the finest gold, which on cooling glittered and shone as gold indeed. A goldsmith to whom he took this declared it to be the purest gold that he had ever seen and offered to buy it at fifty forms the ounce. Amongst others the Master of the Mint came to examine the gold and asked that a small part might be placed at his disposal for examination. Being put through the tests with aqua fortis and antimony it was pronounced pure gold of the finest quality. Helvetius adds in a later part of his writing that there was left in his heart by the Artist a deeply seated conviction that 'through metals and out of metals, purified by highly refined and spiritualized metals, there may be prepared the Living Gold and Quicksilver of the Sages, which bring both metals and human bodies to perfection.'

In the Helvetius tract is also testimony of *Kuffle* and of his conversion to a belief in alchemy as the result of an experiment which he had been able to perform himself, although no indication is given of the source from which he obtained his powder of projection.

Secondly, there is an account of a silversmith named *Gril,* who in the year 1664 at the city of the Hague, converted a pound of lead partly into gold and partly into silver, using a tincture received from a certain John Caspar Knoettner. This projection was made in the presence of many witnesses and Helvetius himself examined the precious metals obtained from the operation.

In 1710 *Sigmund Richter* published his 'Perfect and True Preparation of the Philosophical Stone' under the auspices of the Rosicrucians. Another representative of the Rosy Cross was the mysterious *Lascaris,* a descendant of the royal house of Lascaris, an old Byzantine family, who spread the knowledge of the Hermetic art in Germany during the eighteenth century. Lascaris affirmed that when unbelievers beheld the amazing virtues of the Stone they would no longer be able to regard alchemy as a delusive art. He appears to have performed transmutation in different parts of Germany and then to have disappeared into the blue and so out of history.

CHAPTER VII: ENGLISH ALCHEMISTS

In England the first known alchemist was *Roger Bacon,*[8] a scholar of outstanding attainment, who was born in Somersetshire in 1214. He made extraordinary progress even in his boyhood studies, and on reaching the required age joined the Franciscan Order. From Oxford he passed on to Paris where he studied medicine and mathematics. On his return to England he applied himself to the study of philosophy and languages, with such success that he wrote grammars of the Latin, Greek, and Hebrew tongues.

Although Bacon has been described as a physician rather than a chemist, we are indebted to him for many scientific discoveries. He was almost the only astronomer of his time and in this capacity rectified the Julian calendar which, although submitted to Pope Clement IV in 1267, was not put into practice until a later Papacy. He was responsible also for the physical analysis of convex glasses and lenses, the invention of spectacles and achromatic lenses, and if not for the actual construction, at any rate for the theory of the telescope. As a student of chemistry he called attention to the chemical role played by air in combustion, and having carefully studied the properties of saltpetre, taught its purification by dissolution in water and by crystallisation.

From certain of his letters we may learn that Bacon anticipated most of the achievements of modern science. He maintained that vessels might be

---

[8] See The R.A.M.S. Library of Alchemy Volume 32: Four Works of Roger Bacon.

constructed which would be capable of navigation
without rowers, and which, under the direction of a
single man, could travel through the water at a
speed hitherto undreamt of. He also predicted that
it would be equally possible to construct cars which
'might be set in motion with marvellous rapidity,
independently of horses and other animals,' and
flying machines which would beat the air with
artificial wings.

It is scarcely surprising that in the atmosphere of
superstition and ignorance which reigned in Europe
during the middle ages Bacon's achievements were
attributed to his communication with devils, and
that his fame spread through Western Europe not as a
savant, but as a great magician! His great services
to humanity were met with censure, not gratitude,
and to the Church his teachings seemed particularly
pernicious. She accordingly took her place as one of
his foremost adversaries, and even the friars of his
own order refused his writings a place in their
library. His persecutions culminated in 1279 in
imprisonment and a forced repentance of his labours
in the cause of art and science.

Amongst his many writings there are extant two or
three works on alchemy from which it is quite
evident that not only did he study and practice the
science, but that he obtained his final objective,
the Philosophers' Stone. Doubtless during his
lifetime his persecutions led him to conceal
carefully his practice of the Hermetic art and to
consider the revelation of such matters unfit for
the uninitiated. 'Truth,' he writes, 'ought not to
be shown to every ribald, for then that would become

most vile which, in the hand of a philosopher, is the most precious of all things.'

Sir George Ripley, Canon of Bridlington Cathedral, Yorkshire, placed alchemy on a higher level than many of his contemporaries by dealing with it as a spiritual and not merely a physical manifestation.[9] He maintained that alchemy is concerned with the mode of our spirit's return to God who gave it. He wrote in 1471 his 'Compound of Alchemy' with its dedicatory epistle to Edward IV. It is also reported of this Canon of Bridlington that he provided funds for the Knights of St. John by means of the Philosophers' Stone.

In the sixteenth century *Pierce,* the Black Monk, wrote on the Elixir the following:

'Take earth of Earth, Earth's Mother, Water of Earth, Fire of Earth and Water of the Wood. These are to lie together and then be parted. Alchemical gold is made of three pure souls, purged as crystal. Body, soul, and spirit grown into a Stone, wherein there is no corruption: this is to be cast on Mercury and it shall become most worthy gold.'

Other works of the sixteenth century include *Thomas Charnock's* 'Breviary of Philosophy' and the *additaminta* thereto, and 'Enigma' in 1572. He also wrote a memorandum in which he states that he attained the transmuting powder when his hairs were white.

---

[9] See The R.A.M.S. Library of Alchemy Volume 5: Three Works of George Ripley.

In the sixteenth century also lived *Edward Kelly,* born 1555. He seems to have been an adventurer, and is reputed to have lost his ears at Lancaster on an accusation of producing forged title deeds. Whether this is true or not, the fact remains that *Dr. Dee, a* learned man of the Elizabethan era, was very interested in Kelly's clairvoyant visions, although it is difficult to determine whether Kelly really was a genuine seer since his life was such an extraordinary mixture of good and bad.

In some way or other Kelly does appear to have come into possession of the Red and White Tinctures, since Elias Ashmole printed at the end of 'Theatrum Chemicum Britannicum' a tract entitled 'Sir Edward Kelly's Work' and says:

''Tis generally reported that Doctor Dee and Sir Edward Kelly were so strangely fortunate as to find a very large quantity of the Elixir in some part of the ruins of Glastonbury Abbey, which was so incredibly rich in virtue (being one upon 272,330), that they lost much in making projection by way of trial before they found out the true height of the Medicine.'

How true that may be is a moot point, but it is a fact that in March 1583 the Count Palatine of Siradia, Prince of Poland, Adalbert Alask, while visiting the Court of Queen Elizabeth, sought an acquaintance with Dr. Dee to discuss his experiments, in which he became so interested that he was accompanied by Dee and Kelly and their families on his return to Cracow. The Prince took them from Cracow to Prague in anticipation of favours at the hand of the Emperor, Rudolph II, but their attempt to get into touch with Rudolph was unsuccessful. In Prague at that time a great interest was evinced in alchemy by all and sundry, but in 1586, by reason of an edict of Pope Sixtus V, Dee and Kelly were forced to flee the city.

They finally found peace and plenty at the Castle of Trebona in Bohemia as guests of Count Rosenberg, the Emperor's Viceroy in that country. During that time Kelly made projection of one minim on an ounce and a quarter of mercury and produced nearly an ounce of best gold, which gold was afterwards distributed from the crucible.

In February 1588, following a breach between them, the two men parted, Dee making for England and Kelly for Prague, where Rosenberg had persuaded the Emperor to quash the Papal decree. Through the introduction of Rosenberg, Kelly was received and honoured by Rudolph as one in possession of the Great Secret of Alchemy. From him he received besides a grant of land and the freedom of the city, a councillorship of state and apparently a title, since he was known from that time forward as Sir Edward Kelly. These honours are evidence that Kelly had undoubtedly demonstrated to the Emperor his knowledge of transmutation, but the powder of projection had now diminished, and to the Emperor's command to produce it in ample quantities, he failed to accede, being either unable or unwilling to do so. As a result he was cast into prison at the Castle of Purglitz near Prague where he remained until 1591, when he was restored to favour. He was interned a second time, however, and in 1595, according to chronicles, whilst attempting to escape from his prison, fell from a considerable height and was killed at the age of forty.

In the seventeenth century lived *Eugenius Philalethes* or Thomas Vaughan. Vaughan came from Wales and his writings were regarded as an illustration of the purely spiritual mystery within the science of alchemy, but whatever the various interpretations put upon his work, Vaughan was undoubtedly endeavouring to show that alchemy was demonstratable in every phase of consciousness, physical, mental, and spiritual. His work, 'Lumen de Lumine,' is an alchemical discourse and deals with his subject in the phases I have mentioned. His medicine is a spiritual substance inasmuch as it is the Quintessence or the Divine Life manifesting through all form, both *physical* and spiritual. His gold is the philosophic gold of the physical world as well as the wisdom of the spiritual. His stone is the touchstone which transmutes everything and is again spiritual and physical, and the statement that the Medicine can only be contained in a glass vessel signifies a tangible glass container as well as the purified body of the adept.

Thomas Vaughan was a Magus of the Rosicrucian Order and he knew and understood that the science of alchemy as such must manifest throughout all planes of consciousness.

Eirenaeus Philalethes, by reason of his very numerous writings, must be mentioned. There has been much discussion as to whether this was the name of another adept, or merely another pen name for Vaughan. Mr. Waite has attempted to prove to his satisfaction that they were two different men.

'Personally, I should attribute both names to Thomas Vaughan, but although the question of these authors' identity may make interesting debating material, it is of negligible importance from the standpoint adopted in this book.

In his preface to the Open Entrance from the 'Collectanea Chymica,' published by William Cooper in 1684, he gives testimony:

'I being an adept anonymous, a lover of learning, and a philosopher, decreed to write this little treatise of medicinal, chemical and physical secrets in the year of the world's redemption 1645, in the three and twentieth year of my age, that I may pay my duty to the Sons of Art, that I might appear to other adepts as their brother and equal. Now therefore I presage that not a few will be enlightened by these my labours. These are no fables, but real experiments which I have made and know, as every other adept will conclude by these lines. In truth, many times I laid aside my pen, designing to forbear from writing, being rather willing to have concealed the truth under a mask of envy, but God compelled me to write and Him I could in no wise resist, who alone knows the heart and unto Whom be glory forever. I believe that many in this last age of the world shall be rejoiced with the Great Secret because I have written so faithfully, leaving of my own will nothing in doubt for a young beginner. I know many already who possess it in common with myself, and am persuaded that I shall yet be acquainted in the immediate time to come. May God's most holy will be done therein. I acknowledge myself all unworthy of bringing those things about, but in such matters I submit in adoration to Him, to Whom all creation is subject, Who created all to this end, and having created, preserves them.'

He then goes on to give an account of the transmutation of metals into silver and gold, and also of the fact that the medicine administered to some at the point of death affected their miraculous recovery.

Of one occasion he writes:

'On a time in a foreign country I would have sold so much pure silver worth £600, but although I was dressed like a merchant they said unto me presently that the said metal was made by Art. When I asked their reasons it was answered "We know the silver that comes from England, Spain, and other places, but this is none of these kinds." On hearing this I withdrew suddenly, leaving the silver behind me as well as its price and never returning."

Again he remarks:

'I have made the Stone: I do not possess it by theft but by the gift of God. I have made it and daily have it in my power, having formed it often with my own hands. I write the things that I know.'

In the last chapter of the Open Entrance is his message to those who have attained the goal:

'He who hath once, by the blessing of God, perfectly attained this Art, I know not what in the world he can wish but that he may be free from all snares of wicked men so as to serve God without distraction.

But it would be a vain thing by outward pomp to seek for vulgar applause. Such trifles are not esteemed by those who have this Art, nay, rather they despise them. He therefore whom God hath blessed with this talent has this field of content. First, if he should live a thousand years and every day provide for a thousand men, he could not want, for he may increase his Stone at his pleasure, both in weight and virtue so that if a man would, one man might transmute into perfect gold and silver all the imperfect metals that are in the whole world. Secondly, he may by this Art make precious stones and gems, such as cannot be paralleled in Nature for goodness and greatness. Thirdly and lastly, he hath a Medicine Universal, both for prolonging life and curing of all diseases, so that one true adeptist can easily cure all the sick people in the world I mean his medicine is sufficient.

'Now to the King, Eternal, Immortal and sole Almighty, be everlasting praise for these His unspeakable gifts and invaluable treasures. Whosoever enjoyeth this talent, let him be sure to employ it to the glory of God and the good of his neighbours, lest he be found ungrateful to God his Creditor--who has blessed him with so great a talent--and so be in the last day found guilty of misproving it and so condemned.'

His principal works are 'An Open Entrance to the Shut Palace of the King,' 'Ripley Revived,' 'The Marrow of Alchemy' in verse, 'Metallorum Metamorphosis,' 'Brevis Manuductio ad Rubinem Coelestum,' 'Fone Chemicae Veritatis,' and a few others in the 'Musaeum Hermiticum' and in Manget's collection. There is also the story of a transmutation before Gustavus Adolphus in 1620, the gold of which was coined into medals, bearing the King's effigy with the reverse Mercury and Venus; and of another at Berlin, before the King of Prussia.

Sir Isaac Newton, the famous seventeenth-century mathematician and scientist, though not generally known as an alchemist, was undoubtedly an experimenter in that particular branch of science. If one follows carefully, in the light of alchemical knowledge, the biography of Sir Isaac Newton by J. W. V. Sullivan, I think it is quite easy to realize the experimental theories on which he was working. Sir Arthur Eddington, in reviewing this book, says:

'The science in which Newton seems to have been chiefly interested, and on which he spent most of his time was chemistry. He read widely and made innumerable experiments, entirely without fruit so far as we know.'

His amanuensis records:

'He very rarely went to bed until two or three of the clock, sometimes not till five or six, lying about four or five hours, especially at spring or the fall of the leaf, at which time he used to employ about six weeks in his laboratory, the fire scarce going out night or day. What his aim might be I was unable to penetrate into.'

I think the answer to this might certainly be that Newton's experiments were concerned with nothing more or less than alchemy.

In the same century *Alexander Seton,* a Scot, suffered indescribable torments for his knowledge of the art of transmutation. After practising in his own country he went abroad, where he demonstrated his transmutations before men of good repute and integrity in Holland, Hamburg, Italy, Basle, Strasbourg, Cologne, and Munich. He was finally summoned to appear before the young Elector of Saxony, to whose court he went somewhat reluctantly. The Elector, on receiving proof of the authenticity of his projections, treated him with distinction, convinced that Seton held the secret of boundless wealth. But Seton refused to initiate the Elector into his secret, and was imprisoned in Dresden. As his imprisonment would not shake his purpose he was put to the torture. He was pierced, racked, beaten, seared with fire and molten lead, but still he held his peace. At length he was left in solitary confinement until his release was finally engineered by the adept Sendivogius. Even to his friend he refused to reveal the secret until shortly before his death, two years after his escape from prison, when he presented Sendivogius with his transmuting powder.

CHAPTER VIII: THE COMTE DE ST. GERMAIN

It is rather remarkable that in the history of alchemy the Comte de St. Germain has not been mentioned. There is no doubt that he was an expert in the art, but of the many stories related about this remarkable man, his achievements in this particular sphere seem to play no part.

St. Germain was a baffling personality. As far as can be ascertained he was the son of Prince Racozy of Transylvania, but, in any case, there can be no doubt that he was of noble birth, a man of great culture and refinement. His history as far as it is known is well worth reading, but does not come within the scope of this book, which is solely concerned with his interest in the alchemic art. To those of my readers interested in dietetics, it may be a point of interest that most of his biographers have noted his habits with regard to food. It was diet, he declared, combined with his marvellous elixir, which constituted the true secret of his longevity, for it may be remembered that records of St. Germain's various appearances in Europe extend over a period of 110 years, during which time his appearance never altered. Always he appeared as a well-preserved man of middle age. Madame la Comtesse d'Adhemar, for example, in 'Souvenirs de Marie Antoinette,' gives an excellent description of the Comte, whom Frederick the Great referred to as 'the man who does not die,' and Mrs. Cooper Oakley in her monograph, 'The Comte de St. Germain, the Secret of Kings,' traces him under his various names between the years 1710 and 1822.

The Italian adventurer, Jacques de Casanova de
Seingalt, grudgingly admits that the Comte was an
adept of the magical arts and a skilled chemist.
Upon his telling St. Germain that he was suffering
from an acute disease, the Comte invited Casanova to
remain for treatment, saying that he would prepare
fifteen pills which in three days would restore him
to perfect health.

Of St. Germain's athoeter Casanova writes:

'Then he showed me his magistrum, which he called
Athoeter. It was a white liquid contained in a well
stopped phial. He told me that this liquid was the
universal spirit of Nature and that if the wax of
the stopper was pricked ever so slightly, the whole
contents would disappear. I begged him to make the
experiment. He thereupon gave me the phial and the
pin and I myself pricked the wax, when, lo, the
phial was empty.'

Casanova further records an incident in which St.
Germain changed a twelve sous piece into a pure gold
coin. There is other evidence that the celebrated
Count possessed the alchemical powder by which it is
possible to transmute base metals into gold. He
actually performed this feat on at least two
occasions as stated by the writings of
contemporaries. The Marquis de Valbelle, visiting
St. Germain in his laboratory, found the alchemist
busy with his furnaces. He asked the Marquis for a
silver six-franc piece, and covering it with a black
substance, exposed it to the heat of a small flame
or furnace. M. de Valbelle saw the coin change
colour until it became a bright red. Some minutes
after, when it had cooled a little, the adept took

it out of the cooling vessel and returned it to the Marquis. The piece was no longer silver but of the purest gold. Transmutation had been complete. The Comtesse d'Adhemar had possession of this coin until 1766, when it was stolen from her secretary.

One author tells us that St. Germain always attributed his knowledge of occult chemistry to his sojourn in Asia. In 1755 he went to the East for the second time, and writing to Count von Lamberg he said: 'I am indebted for my knowledge of melting jewels to my second journey to India.'
There are too many authentic cases of metallic transmutations to condemn St. Germain as a charlatan for such a feat. The Leopold Hoffman medal, still in the possession of that family, is the most outstanding example of the transmutation of metals ever recorded. Two-thirds of this medal was transformed into gold by the monk *Wenzel-Seiler,* leaving the balance silver, which was its original state. In the circumstances fraud was impossible as there was but one copy of the medal extant.

For these notes on incidents in St. Germain's life I am indebted to Mr. Manly Hall's introductory material and commentary to the 'Most Holy Trinosophia' (Comte de St. Germain).

The 'Most Holy Trinosophia,' or 'The Most Holy Threefold Wisdom,' is composed of twelve sections. It is at the same time a picture of the process of Initiation and an Alchemical treatise, a fact which careful perusal will establish. Let me quote from Section XII:

'The hall into which I had just entered was
perfectly round it resembled the interior of a globe
composed of hard transparent matter, as crystals, so
that the light entered from all sides. Its lower
part rested upon a vast basin filled with red sand.
A gentle and equable warmth reigned in this circular
enclosure. With astonishment I gazed around this
crystal globe when a new phenomenon excited my
admiration. From the floor of the hall ascended a
gentle vapour, moist and saffron yellow. It
enveloped me, raised me gently and within thirty-six
days it bore me up to the upper part of the globe.
Thereafter the vapour thinned. Little by little I
descended and finally found myself again on the
floor. My robe had changed its colour. It had been
green when I entered the hail, but now changed to a
brilliant red.'

Here is a picture of the pelican in its sand bath,
the process of the sublimation of the contents, and
the change of colour which takes place in one of the
laboratory processes in the preparation of the
Philosophers' Stone. That this preparation is a
physical process carried out in a laboratory with
water, retorts, sand-bath, and furnaces, there is no
doubt. That alchemy is purely a psychic and
spiritual science has no basis in fact. A science to
be a science must be capable of manifestation on
every plane of consciousness; in other words it must
be capable of demonstrating the axiom 'as above, so
below.' Alchemy can withstand this test, for it is,
physically, spiritually, and psychically, a science
manifesting throughout all form and all life.

The various foregoing records should in some measure
bear testimony to the claim of alchemy to be a

physical science based on an inner knowledge of the properties of metals. Casanova's description of St. Germain alone is evidence that as recently as the latter part of the eighteenth century, at any rate, a method of preparing a physical 'Stone,' capable of transmuting metals and curing disease was in practice.

Modern science knows of no substance that can change lead or quicksilver into the likeness of solid gold by the mere addition of a grain of red powder, and may therefore choose to scoff at the alchemists' assertions as products of a too-fertile imagination, at their writings as 'gibberish.' But the fact must be borne in mind that the 'assertions' were corroborated by impartial observers, and that the 'gibberish' of the Hermetic tracts is scarcely less intelligible to the layman than is modern chemical phraseology.

PART II: THEORETICAL

CHAPTER I: THE SEED OF METALS

In this section I am placing before my readers some alchemystical teachings, together with my own interpretation of the theory of alchemy, in an attempt to clarify some of the apparent jargon in which the alchemist expressed his thoughts, and to demonstrate the scientific truth contained therein-- a truth as self-evident and comprehensible as any scientific theory of today.

Instead of dealing with chemistry, occultism, and religion as distinct and separate subjects, alchemy has definitely taught the unity of all Life and Manifestation. It has attempted, and I think successfully, to correlate chemistry, occultism, religion, astrology, magic, and mythology, and to present them all as parts of the One Manifestation. It has attempted also to show that as the health and well-being of the body are as necessary to true religion as true religion is necessary to a healthy and balanced body, so occultism, elucidating as it does the unseen aspects of man, is necessary to both. By true religion, of course, I mean, not the dogmatic teaching of any one church or sect, but the Law of Life and Living; and by occultism, the manifestation of Powers working through and with Man to his ultimate perfection.

That all things proceed from One Thing by the Will of the One Being, that is, that all Manifestation proceeds from one, is the axiom that lies at the root of the theory of all alchemical science. The Hermetic Tract expressed it thus: 'As all things were produced from One by the Mediation of One, so all things are produced from this One Thing by adaptation,' or, in other words, the One in Manifestation has become many. From this One, this Seed, as it were, which the alchemist has called the Alkahest, have proceeded three, Mercury, Sulphur, and Salt, and again from these three have proceeded the many.

Now we must remember that these terms are used by the alchemist very much as the modern chemist uses his terms, which when all is said, convey about as much or as little to the lay mind as do those of the alchemist. The alchemist's Mercury, therefore, must not be confused with the metallic mercury which it resembles neither in texture nor appearance, neither must the Sulphur necessarily possess the qualities of sulphur as we know it, but to a student of alchemy these two substances, together with their salt, convey the idea of the Spirit, the Soul, and the Body. As Paracelsus said: "It is not, however, the common Mercury and the common Sulphur which are the matter of metals, but the Mercury and the Sulphur of the Philosophers are incorporated and inborn in perfect metals and in the forms of them."

It may perhaps simplify matters a little if I give at this point some of the alchemical terms used. The Spirit of Mercury, alternatively called the Quintessence of the Philosophers, Aqua Vitae, Water of Paradise, Azoth, Mercury of the Philosophers, has also on account of its extreme volatility been termed the Eagle, for unless its container be very efficiently sealed, it rises into the air and is lost. Now as I have stated in a previous paragraph, when this Spirit of Mercury or Seed of Metals is divided, from it issue two, the White Mercury and the Sulphur, whose oily tincture, being the golden red of the Sun, has earned for it the name of the Red Lion, the Sun, according to astrology, being in the constellation of Leo the Lion. These two, the White and the Red, are looked upon as the female and male principles, the negative and the positive, Lune the Mother and Sol the Father, or Lune the Queen and Sol the King. This idea of the male and female, or positive and negative elements, is as old as time; take, for example, the following extract from the Chinese, translated by Edward Chalmers Werner:

'Mu Kung, or Tung Wang Kung, the God of the Immortals, was also called I Chun Ming and Yu Huang Chun, the Prince Yu Huang.

'The primitive vapour congealed, remained inactive for a time, and then produced living beings, beginning with the formation of Mu Kung, the purest substance of the Eastern Air, and sovereign of the active male principle (yang) and of all the countries of the East. His palace is in the misty heavens, violet clouds form its dome, blue clouds its walls. Hsien Tung "the Immortal Youth" and Yu nu "the Jade Maiden" are his servants. He keeps the register of all the Immortals, male and female.

'Hsi Wang Mu was formed of the pure quintessence of the Western Air, in the legendary continent of Shin Chou. She is often called the Golden Mother of the Tortoise.

'As Mu Kung, formed of the Eastern Air, is the active principle of the male air, and sovereign of the Eastern Air, so Hsi Wang Mu, born of the Western Air, is the passive or female principle (yin) and sovereign of the Western Air. These two principles, cooperating, engender Heaven and Earth and all the beings of the universe, and of the subsistence of all that exists.'

At this point, too, I should explain that the metals have been recognized as the manifestation of planetary influences and named in accordance. Thus

Gold	is termed the	Sun
Silver	" "	Moon
Mercury	" "	Mercury
Tin	" "	Jupiter
Iron	" "	Mars
Copper	" "	Venus
Lead	" "	Saturn

According to this teaching the metal is formed as the result of certain stellar vibrations or waves of energy and consequently carries the characteristic of the planet by which it is influenced. Thus: Gold is the manifestation of the perfect metal even as the Sun is the manifestation of Life on this planet:

Silver, the colour of white, is the Moon, the negative aspect of the Sun:

Mercury, as the planet Mercury, is of a volatile nature, its surface being in constant movement:

Iron is strength and force, Mars being the planet of
energy and force:

Copper is Venus, closely approaching the colour of
gold, Venus being the planet of beauty, and of love:
Lead is Saturn the Tester, cold, and known in
cabbalistic teachings as the root of metals:
Tin is Jupiter, the planet of benevolence and
opulence.

All metals are in a constant state of progression.
By this I mean that Gold, the perfect metal, stands
at the head, the summit of perfection, as it were,
whilst

Gold

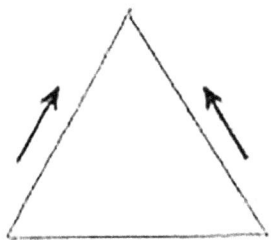

all other metals are on the way towards eventually
becoming gold; thus the alchemist merely does by art
what nature does slowly through the years. Species,
says Friar Bacon, are not transmuted, but rather
their subject matter. It is the subject matter of
the metals, the radical moisture of which they are
uniformly composed, that the alchemist maintains may
be withdrawn by art and transported from inferior
forms, being set free by the force of a superior
ferment or attraction.

Metals have always been recognized by the alchemists
as living, breathing substances, each one having as
its component parts Mercury, Sulphur, and Salt, the

difference in the consistency and characteristics of the metal being due to the proportion of these three principles one to the other.

To illustrate this point, let me quote from Basilius Valentinus, one of the greatest alchemists of the fifteenth century:

'Therefore the metal of Mars (Iron) is found to have the least portion of Mercury, but more of Sulphur and Salt.

'The reader must moreover know concerning the generation of copper, and observe that it is generated of much Sulphur, but its Mercury and Salt are in an equality....

'Among all metals Gold hath the pre-eminence because the sidereal and elementary operation hath digested and refined the Mercury in this Metal the more perfectly to a sufficient ripeness...

'Good Jupiter (Tin) possesses almost the middle or mean place between metals, it being not too hot, nor too cold, nor too warm, nor too moist, it hath no excess of Mercury, nor of Salt, and it hath the least of Sulphur in it....

'I tell thee that Saturn is generated of little Sulphur, little Salt, and much unripe gross Mercury, which Mercury is to be esteemed a froth that floats upon the Water in comparison of that Mercury which is found in Sol (Gold).'

These quotations will illustrate what I intend to convey by my reference to the proportionate relationships of the three substances.

To revert to the subject of the seed of metals, from the 'Speculum' of Arnaud de Villeneuve come these words: 'There is in Nature a certain fine essence, which being discovered and brought by art to perfection converts to itself proportionately all imperfect bodies that it touches,' so that the first matter of all metals and substances is a fixed something altered by the diversities of place, temperature, etc. This 'Essence' has always been recognized by alchemists as the Seed of Metals.

To illustrate my meaning in regard to the Seed of the Species, I quote the following from 'Ether and Reality,' by Sir Oliver Lodge (Messrs. Hodder & Stoughton):

'Matter exists not only in the organic forms of solids, liquids and gases and in the disintegrated forms of electrons and protons, it exists also as the complex molecules known as protoplasm, which for some reason or other has shewn itself to be the vehicle of life. Some forms of matter are endowed with or animated by life. This property of animation is a great mystery; we do not know what Life is, we only see what it can do. We perceive that it can enter into relation with matter, that it has a character and identity of its own, and that it builds up matter to correspond with or to represent identity. Life can take a variety of forms, and every form is characterized by a certain shape; the life of an oak is transmitted to an oak, the life of an elm to an elm. "To every seed his own body." One form of life takes the shape of a bud, another of a fish, another of a quadruped. The varieties of life are innumerable, and are studied in the great science of biology.

'Consider any piece of matter. . . . Contemplate any solid object; a vase, it may be, or a jewel, or a statue; what is it that holds the atoms together in that particular shape? If the atoms were not connected they would be moving about at random, like the atoms of a gas; but they are connected, crystallized as it were, together by the forces of cohesion. Even in a liquid they are held together into a body of definite size, though not a definite shape; a liquid has size though not shape; a gas has neither; a solid has both. The shape is most definite and law-abiding in a crystal; but in a plant or animal it has a definite character too--not so definite as in a crystal, a good deal of variety is possible, yet an animal or vegetable body has an undoubted character of its own, even to minute detail. And this character is handed down from one generation to another, modified perhaps, but only slowly, by the age-long process of Evolution.'
This extract from Sir Oliver Lodge I have quoted in full, for in the words 'to every seed his own body' lies the whole doctrine of alchemy, which has recognized a metallic seed peculiar to all metals.

CHAPTER II: THE SPIRIT OF MERCURY

In the previous chapter I spoke of the substances Mercury, Sulphur, and Salt as being analagous to the Spirit, Soul, and Body. What I intend to convey is that the Spirit of the Metal is the Spirit of Mercury (a volatile essence which in its gaseous state is an Aether), the Sulphur is the Soul or the Blood, and the Salt the Ashes or the Body.

Again I quote from Basilius Valentinus, Father of Modern Chemistry:

'*Of the Spirit of Mercury*.'

'Though I have a peculiar Stile in writing, which will seem strange unto many, causing strange Thoughts and Fancies in their Brains, yet there is reason enough for my so doing; I say enough, that I may remain by my own experience, not esteeming much of others prating, because it is concealed in my knowledge, Seeing having alwaies the preheminence before Hearing, and Reason hath the praise before Folly: Wherefore I now say, that all visible, tangible things are made of the Spirit of Mercury, which excels all earthly things of the whole world, all things being made out of it, having their Off-spring only from it; for all is found therein which can perform all whatsoever the Artist desires to find; It is the beginning to operate Metals, when it is become a spiritual Essence, which is meer Air flying to and fro without wings; it is a moving wind, which after it is expelled its dwelling by *Vulcan,* it is driven into its *Chaos,* where it again enters, and resolves itself into the Elements, where it is elevated and attracted by the Sydereal Stars after a Magnetical manner unto themselves, out of love, whence he proceeded before, and was operated because it affects its like again, and attracts it to it. But if this Spirit of *Mercury* can be caught, and made corporal, it resolves into a Body, and becomes a pure, clear, transparent water, which is the true spiritual water, and the first *Mercurial* Root of the Minerals and Metals, spiritual, intangible, incombustible, without any mixture of earthly Aquosity; it is that Celestial water, whereof very much hath been written; for by this Spirit of *Mercury* all Metals, may if need require, be broken, opened, and resolved into their first Matter, without Corrosive; it renews the age of Man or Beast, even as the Eagles; it consumes all evil, and conducts a long Age to long Life. This Spirit of *Mercury* is the Master-Key of my Second Key, whereof I wrote in the beginning; wherefore I will call; *Come ye Blessed of the Lord, be anointed, and refreshed with water, and embalm your Bodies, that they may not putrefie or stink;* for this Celestial Water is the beginning, the Oyl, and the means, seeing it burns not, because it is made of spiritual Sulphur; the Salt Balsam is corporal, which is united with the Water by the Oyl, whereof I will afterwards treat more at large, when I shall write of them, and mention them.

'And that I may further declare what is the Essence, Matter and Form of the Spirit of Mercury, I say, that its Essence is blessed, its Matter spiritual, and its Form earthly, which yet must be understood by an incomprehensible way; these are indeed harsh Expressions, many will think, thy Proposals are all vain, strange Effusions, raising wonderful

Imaginations, and true it is that they are strange, and require strange people to understand these Sayings; it is not written for Peasants, how they should grease Cart-wheels, nor is it written unto those who have no knowledge of the Art, though they be never so learned, or think themselves so; for I only account them Learned, who next unto Gods Word, learn to know Earthly things, which must be pondered and judged by the Understanding, founded upon a true Knowledge, to distinguish Light from Darkness, who choose that which is good, and reject the evil.

'It is needless for you to know what the beginning of this Spirit of *Mercury* requires, because it can in no wise help nor advantage you, only take notice of this, that its beginning is supernatural, out of the Celestial, Sydereal and Elementary, bestowed on it from the beginning of the first Creation, that it may enter further into an Earthly Substance. But because this is necessary which hath been declared to you, leave the Celestial to the Soul, apprehend it by Faith, and let the Sydereal likewise alone, because these Sydereal Impressions are invisible and intangible, the Elements have already brought forth the Spirit perfect into the world by the Nutriment, therefore let that alone likewise; for man cannot make the Elements, but only the Creator, and remain by thy made Spirit which is already formal and unformal, tangible and intangible, and yet is presented visibly. So have you enough of the first Matter, out of which all Metals and Minerals grow, and is one only thing, and such a matter which unites itself with the *Sulphur* in the following Chapter, and enters into a Coagulation with the *Salt* of the first Chapter, that it may be one Body, and a perfect Medicine of all Metals, not only to bring forth in the Earth at the beginning, as in the great World, but also by help of the vaporous Body to transmute and change, together with the augmentation in the lesser World: Let not this seem strange to you, seeing the Most High hath permitted, and Nature undertaken it.

'Many will not believe this, esteeming it impossible, despise and vilify these Mysteries, which they understand not in the least, they may remain Fools and Idiots till an illumination follows, which cannot be without God's Will, but remains till the time predestinate. But wise and discreet men, who have truly shed the sweat of their Brows, will be my sufficient witnesses, and confirm the Truth, and indeed believe and hold for a truth all that which I write in this case, as true as Heaven and Hell are preordained, and proposed as Rewards of good and evil to the Elect and Reprobate. Now I write not only with my hands, but my Mind, Will and heart constrain me to it: Those who are highly conceited, illuminated, and world-wise, hate, envy, scandalize, defame and persecute this Mystery to the utmost Rind, or innermost Kernel, which hath its beginning out of the Center; but I know assuredly, there will come a time, when my Marrow is wasted, and my Bones dried up, that some will take my part heartily, after I am in the Pit; and if God would permit it, they would willingly raise me from the dead; but that cannot be; wherefore I have left them my Writings, that their Faith and Hope may have a Seal of Certainty and Truth, to testifie of me what my last Will and Testament was, which I ordained for the poor, and all the Lovers of Mysteries, though it did not behove me to have wrote so much, yet I could not refrain without prejudice to my Soul, but to drive a Light or Flash through a Cloud, that the Day might be observed, and the dark Night, thick and gloomy, rainy Weather expelled.

'Now how the *Archaeus* operates further by the Spirit of Mercury in the Earth, or Veins of the Earth, take this Advice, that after the spiritual Seed is formed by the impression of the Stars from above, and fed by the Elements, it is a Seed, and turns itself into a *Mercurial* Water, as first of all the great World was made of nothing, for when the Spirit moved upon the Water, the Celestial Heat must needs raise a Life in the cold waterish and earthly Creatures; in the great World it was Gods Power, and the Operation of the Celestial Lights; in the little World it is likewise Gods Power, and the Operation to work into the Earth by his Divine and Holy Breath. Moreover the Almighty gave and Ordained means to accomplish it, that one Creature had obtained power to operate in the other, and the one to help and assist the other, to perform and fulfil all the Works of the Lord; and so an influence was permitted the Earth to bring forth by the Lights of Heaven, as also an internal Heat, to warm and digest that which was too cold for the Earth, by reason of its humidity, as unto every Creature a peculiar fashion according to its kind; so that a subtile sulphurous Vapour is stirred up by the Starry Heaven, not the common, but another more clarified and pure Vapour, distinct from others, which unites itself with the *Mercurial* Substance; by whose warm property, in process of time, the superfluous Moisture is dryed up, and then when the soulish property comes to it, which gives a preservation to the Body and Balsam, operating first into the Earth by a spiritual and sydereal influence, then are Metals generated of it, as it

pleaseth the Mixture of the three Principles, the
Body being formed according as it assumes unto it
the greatest part of those three. But if the Spirit
of *Mercury* be intended and qualified from above upon
Animals, it becomes an Animal Substance; if it goes
upon Vegetables by order, it becomes a Vegetable
Work; but if, by reason of its infused nature, it
fall on Minerals, it becomes Minerals and Metals,
yet each one hath its distinction as they are
wrought, the Animals for themselves, the Vegetables
on another manner and form by themselves, and so
likewise the Minerals, each one a several way,
whereof to write particularly would be too tedious,
and yield large and Various Narrations.

'This is the summe in brief, that without the Spirit
of *Mercury,* which is the only true Key, you can
never make Corporal Gold potable, nor the
Philosophers' Stone. Let it remain by this
Conclusion, be silent; for I myself will at present
say no more, because Silence is enjoyned thee and me
by the orderly Judge, recommending the Execution and
further search thereof to another, who hath not as
yet reduced the Matter into a right Order.'

And here the words of Alexander von Suchten, from
the 'Blessed Casket of Nature's Marvels' by
Benedictus Figulus:

'The primary matter of man and the primary matter of
the great world are one and the same thing. But this
primary matter of the world and of man is a
Crystalline Water of which Holy Writ says "Before
God created Heaven and Earth, the Spirit of the Lord
brooded over the waters." This water became a
primary matter of both. But where remains the Spirit

of the Lord, which brooded over the waters, after the two worlds, i.e. heaven and earth, and man had been created from the same? I reply, in the primary matter of man and of the world, God who is Perfection, has wished to dwell in Man. But here the following question might be put; how did man know-- since the primary matter of man and the world is a crystalline water--how could man know whether the Spirit of the Lord had remained in this primary matter of the world, or of man? I reply, he knew it by the Art of Water, for Water was his teacher. This teacher shewed him how the world dies, how the Spirit departs from it, how the body is without spirit, the spirit without body. He saw how the spirit returns to the body, and the body revives. He saw by the decay of the world that it did not become again what it had been before. Hence it became plain to him that God dwells not in that which passes away, but in that which is eternal."

CHAPTER III: THE QUINTESSENCE (I)

Space, whether inter-planetary, inter-material, or inter-organic, is filled with a subtle fluid or gas, which we call, as did the ancients, Aith-in-Solintaire Aether. This fluid or gas, unchangeable in composition, indestructible, invisible, pervades everything and all matter. Metal, mineral, tree, plant, animal, man; each is charged with the Ether in varying degrees. All life on the planet is charged in like manner; a world is built up in this fluid, and moves through a sea of it.

Ether, which the occultist terms astral light, determines the constitution of bodies. Hardness and softness, solidity and liquidity, all depend on the relative proportion of ethereal and ponderable matter of which they are composed.

The arbitrary division and classification of physical science, the whole range of physical phenomena, proceeds from the Primary Aether, for Science has reduced matter as we know it to Ether, which, although not solid matter, is still matter. When most of us speak of matter, of course, we usually visualize solid substance, but it has been proved by Science that matter is not actually solid, but merely a stress, a strain in the Ether. The atoms, and finer still, the electrons and protons of which it is composed, all move in a sea of Ether, so that in accordance with this theory, the very air we breathe, the very bodies we inhabit, all must likewise be moving in this sea of Ether, the parent element from which all manifestation has come.

This principle that all things proceed from one is demonstrable in the physical; in the principles of Biology, the multicellular organisms, complex as they may be in their structure, nevertheless arise from a single cell. Science postulates that all matter is composed of atoms: the atoms, however, are composed of protons and electrons, and the electrons in their turn are evidently composed of Ether. This Ether is a universal connecting medium filling all Space to the furthest limits, penetrating the interstices of the atoms without a break in its continuity, and so completely does it fill Space that it is sometimes identified with Space, and has, in fact, been spoken of as Absolute Space.

'The Ether of Space,' according to Sir Oliver Lodge, 'is a theme of unknown and apparently infinite magnitude, and of a reality beyond the present conception of man. It is that of which everyday material consists, a link between the worlds, a consummate substance of overpowering grandeur. By a kind of instinct one feels it to be the home of spiritual existence, the realm of the awe-inspiring and supernal. It is co-extensive with the physical universe, and is absent from no part of space. Beyond the furthest star it extends, in the heart of the atom it has its being. It permeates and controls and dominates all. It eludes the human senses and can only be envisaged by the powers of the mind.

'Yet the Ether is a physical thing; it is not a physical entity, it has definite properties. It is not matter any more than hydrogen and oxygen are water, but it is the vehicle of both matter and spirit. . . '

Now the occultist has divided matter, seen and unseen, into seven principles or planes, and of these the fifth principle, or Quintessence, corresponds to Science's Ether of Space. If we are willing to admit that there is truth in this statement, then we may begin to see that alchemy is based on absolute Law. All the forces of our scientists have originated in the Vital Principle, that one Collective Life of our Solar System, which life is a part of, or rather one of the aspects of, the One Universal Life.

During life there is present in man a finely diffused form of matter, a vapour filling not merely every part of his physical body, but actually stored in some parts; a matter constantly renewed by the vital chemistry, a matter as easily disposed of as the breath, once the breath has served its purpose. Of this matter Paracelsus wrote:

'The Archaeus is an essence that is equally distributed in all parts of the human body. . . . The Spiritus Vitae takes its origin from the Spiritus Mundi. Being an emanation of the latter, it contains the elements of all cosmic influences, and is therefore the cause by which the action of the cosmic forces act upon the body of man.'

This Archaeus is of a magnetic nature and is not enclosed in a body but radiates within and around it like a luminous sphere. Alchemy and alchemy alone, within historical period, and in so-called civilised countries, has succeeded in obtaining a real element, or a particle of homogeneous matter, the Mysterium Magnum of Paracelsus. By his age-old science the alchemist may set free this Vital

Principle in his laboratory, destroy the body of the metal on which he is working, purify its salt, and bring its principles together in a higher form. This process, which is after all but a miniature reproduction of a superior process in operation around us all the time, undoubtedly proceeds from Master Intelligences who have lived at some time or another on this Earth.

It is a pity that Science must always reject old ideas and cast them away as useless before rediscovering them as something new to be incorporated in its current theories. To discard the alchemist's theories is as intelligent as to dismiss as rubbish Einstein's Theory of Relativity merely because one does not happen to understand his language. Some of our scientific men have realized this, for F. Hoefer in 'Histoire de la Chimie' (Paris 1866) remarks: 'The systems which confront the intelligence remain basically unchanged through the ages, although they assume different forms. Thus, through mistaking form for basis, one conceives an unfavourable opinion of the sequence. We must remember that there is nothing so disastrous in Science as the arrogant dogmatism which despises the past and admires nothing but the present.'

If Science would but try to understand the conception of the Universe as taught by occultism throughout the ages, taking as its starting-point the teaching of the One Life in Manifestation, its seven planes of consciousness, its infinite forces, and as the basis of its philosophy the Hermetic axiom 'as above, so below,' it would found a system based on eternal Truth instead of on a quicksand of theories. Science will never really understand the truth about life until it reaches this realization, which cannot be attained through its instruments and appliances, but only through the inner powers of the mind.

THE QUINTESSENCE. (II)

'Nothing of true value is located in the body of a substance, but in the virtue thereof, and this is the principle of the Quintessence, which reduces, say 20 lbs. of a given substance into a single ounce, and that ounce far exceeds the 20 lbs. in potency. Hence the less there is of body, the more in proportion is the virtue thereof.'

Paracelsus has said:

'The Magi in their wisdom asserted that all creatures might be brought to one unified substance, which substance they affirm, may by purification and purgation, attain to so high a degree of subtlety, such divine nature and occult property, as to work wonderful results. For they considered that by returning to the earth, and by a supreme and magical separation, a certain perfect substance would come forth, which is at length, by many industrious and prolonged preparations, exalted and raised up above the range of vegetable substances into mineral, above mineral into metallic, and above perfect metallic substances into a perpetual and divine Quintessence, including in itself the essence of all celestial and terrestrial creatures.'

By this Quintessence or *quintum esse,* Paracelsus meant the nucleus of the essences and properties of all things in the universal world.

From the 'Golden Casket' of Benedictus Figulus comes the following:

'For the elements and their compounds in addition to crass matter, are composed of a subtle substance, or intrinsic radical humidity, diffused through the elemental parts, simple and wholly incorruptible, long preserving the things themselves in vigour, and called the Spirit of the World, proceeding from the Soul of the World, the one certain Life filling and fathoming all things, so that from the three genera, or creatures, Intellectual, Celestial and Corruptible, there is formed the One Machine of the Whole World. This spirit by its virtue fecundates all subjects natural and artificial, pouring into them those hidden properties which we have been wont to call the Fifth Essence, or Quintessence. . . .

But this is the root of life, i.e., the Fifth Essence, created by the Almighty for the preservation of the four qualities of the human body, even as Heaven is for the preservation of the Universe. Therefore is this Fifth Essence and Spiritual Medicine, which is of Nature and the Heart of Heaven, and not of a mortal and corrupt quality, indeed possible. The Fount of Medicine, the preservation of Life, the restoration of Health, and in this may be cherished the renewal of lost youth and serene health be found.'

Turning from the words of the alchemists of the fifteenth and sixteenth centuries to those of a

twentieth century scientist, let me quote from Sir
Oliver Lodge's 'Ether and Reality' once again:

'Apollonius of Tyana is said to have asked the
Brahmins of what they supposed the Cosmos to be
composed.

'"Of the five elements."

'"How can there be a fifth," demanded Apollonius,
"beside water and air and earth and fire?"

'"There is the ether," replied the Brahmin, "which
we must regard as the element of which the gods are
made; for just as all mortal creatures inhale the
air, so do immortal and divine natures inhale the
ether."'

And:

'What you choose to call this unifying "Something"
is of no consequence. The Ancients sometimes spoke
of the "Ether," possibly as an addition to the usual
four elements, and Sir Isaac Newton adopted this
term for the connecting medium. The optical medium
connects the particles together in a solid or a
liquid, and the same medium connects the heavenly
bodies together into systems and clusters and
constellations and nebulae and Milky Way.

'All pieces of matter and all particles are
connected together by the Ether and by nothing else.
In it they move freely, and of it they may be
composed. We must study the kind of connexion
between matter and Ether.

'The particles embedded in the Ether are not independent of it, they are closely connected with it, it is probable that they are formed out of it: they are not like grains of sand suspended in water, they seem more like minute crystals formed in a mother liquor. . .'

Again:

'Speculatively and intuitively we feel to be more in direct touch with the ether than with matter. How we can act on matter is a mystery. How we have constructed and how we move our bodies, we do not know. We are apt to identify ourselves with our bodies. But there is evidence which shows that we are really independent, that we continue in existence, and can leave our bodies behind. Matter is not part of our real being, not of our essential nature it is but an instrument that we use for a time and then discard. Probably we do not act directly upon matter at all. Our will, our mind, our psychic life, probably act directly upon the Ether; and only through it, indirectly, on Matter. Ether is our real primary and permanent instrument. It is in connexion with the Ether that our real being consists; and through it we are able to manipulate the atoms of matter, to move them, to rearrange them, and thus 'employ them to express our thoughts and feelings and to manifest ourselves to other individual entities who in the long course of evolution have been enabled to construct and employ similar most ingenious, though imperfect, instruments of manifestation. By this means we can become aware of a multitude of existences, the whole animal and vegetable kingdom, of which otherwise we might have remained ignorant; by this means our conceptions of existence have been enlarged and extended, the possibilities of friendship enhanced, the perception of a new realm of law and order attained. And thus is our own nature enriched by the effort and experiences belonging to a new and most interesting-- though from our point of view imperfect and rebellious--physical mode of existence.'

And his closing words:

'It is the primary instrument of Mind, the vehicle of Soul, the habitation of Spirit. Truly it may be called the Living Garment of God.'

This comparison between the writings of scientists of different centuries is interesting, since it seems to me that while there may be some difference in actual verbal expression, each man refers to the same principle.

CHAPTER IV: THE QUINTESSENCE IN DAILY LIFE

Since it is not possible for everyone to follow its reactions in the laboratory, I am devoting this chapter to the manifestation of the Quintessence in everyday life, for it is not merely in the laboratory that this vital principle evinces itself, but through all phases and conditions of existence.

Vitamines.

First, what of our food? The physicist has found that for a food to be really worthy of that name it must contain a certain vital essence, which he has called the Vitamine. Without this vital quality, which I believe to be this same Quintessence or Divine Energy, any type of food whatsoever is just so much dead matter. For instance, expeditions on which the men have subsisted entirely on a diet of tinned food have invariably shown that whilst ingesting the bulk of food necessary for the satisfaction of their hunger, they yet suffered from starvation since that food was devoid of its vital principle--the Quintessence or Vitamine. Most of us have read at some time or another of the sufferings of the early navigators who would sail for weeks without sighting land, living the while on dried food. From those islands which could provide anything in the way of fresh meat and fruit they would replenish their miserable stores, and for a time whilst these fresh provisions lasted, the crew would improve in health and vitality, but with the exhaustion of the supply would come depletion of vitality, scurvy, and other trials occasioned by a deficiency diet. Citrous fruits, in particular, were found to be extremely effective for combating scurvy, and British sailors at one time in their history were called 'limies' by reason of the citrous fruit included in their food quota.

This food problem, then, which we have confronting us is surely a proposition of vast dimensions. From all sides we are bombarded with demands for a fitter people, for an A 1 nation, but if this high standard of national health is to be attained, then the food problem of the people must be tackled in all seriousness. While the peoples of the world depend for their sustenance (as the greater part of our Western civilization does today) on a diet of highly refined food, from which all real food value has been extracted in the process of refinement, there is little hope of any improvement in their physical status, and this lack of vitally charged food may easily be a reason, and a very important reason, for such diseases as cancer and kindred complaints; infantile paralysis, sleepy sickness, and influenza. As a preventative to many diseases, medical men are now recommending Vitamin D, but actually this question of Vitamines is only touching the edge of a problem which is of very real importance and urgency to each one of us--the necessity for a diet incorporating in its constituents that vital energy or quintessence without which a food is no food at all.

Digestion.

From the food itself let us turn our attention to the digestion of that food in the human body. In the process of digestion we find a much more complex action taking place than physiology has so far been able to demonstrate. The process of ingesting food into the human stomach is really a mild form of poisoning, and in order to utilize to the best advantage the foodstuffs he is taking, the human being must transmute those foodstuffs, provided for him by the animal and vegetable kingdoms, into a form that the cells of his body can readily take up and assimilate. Without this process of change in digestion, man would probably die of poisoning! For an example of this changing process, take albumen. Albumen in the process of digestion is split up into its amino acids and then brought together again as a human albumen capable of absorption and assimilation by the cells of the human body.

Can any physiologist explain how this change takes place? Physiologically there is no explanation which would elucidate this process, but that it does take place is a fact. In its enactment we have an instance of transmutation, of man taking into his body a lower form of life for its transmutation into something higher, and what is that but an alchemical process? The transmutation of a lower substance into a higher, when it takes place in the body of man, is definitely a function of the unconscious part of the mind--a function not consciously performed by the ordinary individual owing to the fact that the Mind of Man, in the process of building form from the Amoeba upwards, has relegated such functions to the unconscious or subconscious part of the mind, leaving the surface consciousness to carry on with outside problems. Thus whilst all this work of digestion, circulation, breathing, etc., is being carried on by the deeper strata of the mind, the upper strata are free, as I have just said, to deal with the demands of everyday life. How many of us realize, I wonder, that here in this very process of digestion is taking place an act of magic which the average man cannot understand, complacently though he accepts it. Occultists have taught that this process of the transmutation of food in the human body can be helped by the conscious part of the mind (by what some schools would call auto-suggestion). Thus we have an example of man as the medium through which a transmutation of a lower form of matter into a higher may take place.

Breathing.

To take another function of the human body--that of breathing. What has physiology to tell us of the process of breathing? We are taught that the most important function of breathing is the taking of oxygen into the lungs to revivify that venous blood which has lost its oxygen in its circulation of the body, and has to be replaced before it can pass on into the arterial circulation once more.
This is one function of breath, but another, which physiology has so far not touched, is the breathing in of the natural electricity or Vital Principle (the Quintessence) in the atmosphere, which the human body uses as nervous energy. Here again the unseen alchemist is at work, engaged in the absorption of the air around him and its transmutation into something higher for the work in his own body.

This question of breathing brings another in its train--the question of:

The Heart's Action.

Is the heart, as physiology states, an instrument for the pumping of blood through the blood-vessels of the body? Impossible; it would require a much larger and more powerful organ than the heart to pump blood through some of the tiny blood-vessels in the body. The heart is the regulator of the flow, not the pump, the circulation of the body being an electrical process, with the arteries as the positive and the veins as the negative charges. The venous blood being negative is drawn to the lungs which are positive, and there re-charged with the air intaken by the lungs. After receiving its positive charge the blood is repelled from the lungs (since two positive charges repel one another) and flows through the heart to the Aorta, the rate of its flow being regulated by the heart's beat. The Aorta divides and sub-divides throughout the body, giving up its charge to the nervous system, which passage causes the blood once again to become negative, and necessitates its return (through the veins) to the lungs for re-charging. In these days of knowledge of electricity and magnetism, it is only logical to conclude that these so-called mechanical actions of the organs of the body are electrical.

The atom of oxygen is like a sponge that holds a certain amount of etheric force or electricity (the Quintessence), each atom enclosing within itself a charge of vital energy. The human body is a chemical laboratory and the so-called atoms of oxygen, hydrogen, nitrogen, etc., contain within themselves charges of Vital Energy. The Yogi, in describing his breathing exercises, speaks of a certain vital principle of energy which he calls 'Prana,' which is in actual fact another instance of the manifestation of the Quintessence. In his system of breathing the mind is so centered on the act of breathing that this Quintessence of the air is consciously taken in for the revitalization of every part of his body. When you take a holiday in the mountains or by the sea, with beneficial results, the real benefit obtained is from this Quintessence or Vital Energy in the air which you breathe in.

The alchemist, by his laboratory process, is taking this Quintessence or Vital Energy from metals, since he has found in his experience that it is obtained from minerals and metals in a more perfect form than from plant life, the minerals being of the first manifestation.

PART III

CHAPTER I: THE MEDICINE FROM METALS

In our treatment of the human body we have to remember that in composition it is not an inanimate object capable of sustaining the kind of treatment accorded to a sack of sand, but a delicate organism possessed of the capacity of feeling, consciously and unconsciously, and must be handled accordingly. The cell life of the body is selective in the finest sense, the cells rejecting any substance unfit for their use, and consequently it is as reasonable to expect to run a modern aeroplane engine on inferior fuel as to ingest into the human body for its maintenance a drug of a gross nature, or a food devoid of its natural vital principle.

We all have constant proof of the fact that at a certain stage in his life man's body apparently begins to deteriorate, the reason given for this deterioration being the slowing down of the cell activity with the result that the body's wasting process proceeds more rapidly than does the repairing process. This explanation is correct, for as man gets older, the vital energy does not flow through the cells of his body so efficiently as in his youth, and the cells of the body, when unable to obtain their requisite elements, become sluggish in their action and ultimately diseased.

In this connection our ideas on so-called diseased bacteria have to be very much revised; the so-called bacteria is the medium through which the vital energy is transferred to the cell life. This is its work, the purpose for which it was created, and if for some reason the flow of that energy is impeded in its passage, then the bacteria takes its energy from the cell, and at once becomes pathological. For this reason it has been regarded by the medical faculty as the *cause* of the disease; but any bacteriologist will realize how nearly he has approached to the truth of this statement when he in his turn states, for example, that certain types of bacteria are oxygen-eating, that is, in the event of their being unable to obtain their oxygen from such a substance as sugar, they take it from the human body and so debilitate and disease that body. For this reason, if we really desire to become an A 1 race, we must find and understand the preparation of those elements which the human body's cell-life requires to assist its correct functioning, for when the cell-life of the body fails, then the body itself fails also.

Alchemy, as demonstrated by two of its most prominent exponents, Basil Valentine and Paracelsus, is concerned not only with the attainment of the Philosopher's Stone, but with the preparation of medicines, by which is meant the separation of the ethereal from the gross, the true secret of the Spagyric Art.

At the present day we have two definite systems of medicine, the one termed allopathic, the other homeopathy. Both these systems have countless remedies, but neither is by any means perfect, for where the allopath gains his cures, the homeopathist has to admit defeat, and where the homeopathist succeeds, the allopath may fail. The allopath, whose methods are the more widely practised at the moment, maintains that the homeopathist gains his successes through the imagination of his patients, but the homeopathist believes his methods to be the more scientific, since he deals with a more finely divided and spiritualized medium; for while the allopath uses his drugs without trituration, the homeopathist triturates his drugs from the first decimal to the higher potencies even up to the two-hundredth decimal. Even so, although his method is the more perfect of the two, it is still far from the ideal.

The homeopathists, of course, teach that the founder of their system was Hahneman, but in actual fact this is inaccurate. Hahneman merely rediscovered in part a system which had been taught in alchemy for hundreds of years. I say in part because the alchemist's interpretation of the system was very much more perfect than is the modern homeopathist's.

In regard to the question of potencies, I will repeat once again the definition of the Quintessence: 'Nothing of true value is located in the body of a substance but in the virtue thereof. And this is the principle of the quintessence, which reduces, say, twenty pounds to a single ounce, but that ounce far exceeds in potency the entire twenty pounds.' Thus to find the Quintessence of Iron, for example, the metal is changed into its vitriol or salts, which in turn are purified by several washings in distilled water, and after each washing re-crystallised. The salt is then calcined to redness and its spirit drawn off in a special manner and also in its turn carefully distilled several times, the result being a red oil of iron which is its *true* essence, a few drops constituting a dose.

The first essential of a really effective healing agent is that it should contain the Quintessence or vital principle of the herb or metal used, and it is the homeopathist's failure to provide this element in his preparations which entails the loss of the real value of his medicaments.

The allopath's failures lie in the fact that his remedies are always administered in too crude a form. In the administration of a metal, for instance, it must be understood that the body of a metal is worthless, as a medicine, it cannot heal: it is the *essence alone* that is curative. Only too often the body is poisonous, and until that gross part of the metal be broken up, its administration is definitely *harmful*. Probably one of the most common forms of metallic poisoning is that of mercury, but remove the harmful parts of the metal and the healing essence is free to do its work thoroughly. Nitrate of silver is a caustic poison, but remove the gross part of the metal and the essence of the silver is a cure for diseases of the brain. Lead salts are poisonous, it is true, and in many cases their administration has resulted in death from lead poisoning, but remove that poisonous matter and the remaining essence, which is clear, sweet-smelling, and aromatic in taste, forms a cure for all diseases of the spleen. Copper, when the gross body of the metal is removed and the essence unlocked, is invaluable for the nervous system and the kidneys; likewise, tin for the liver, iron for all inflammatory diseases, and the bile, and gold for the heart and general circulation. But gold, too, is only suitable for a medicine when the salts of gold are reduced into the oil of gold and distilled into a golden liquid; then and only then is gold tolerated and utilized by the human body. The salts of gold used at the present day can never be assimilated, for by their present method of preparation they can never be properly distilled and purified.

From the foregoing paragraphs it will be seen that the whole principle of cure rests on the proper separation of this Quintessence to which alchemy, and alchemy alone, provides the key. The whole principle of the system is that the body of the metal impedes the action of the essence, and those metals which have hitherto been regarded as poisonous (mercury, antimony, lead, arsenic) are all non-poisonous and capable of greater curative potency when this process has been faithfully carried out.

A third system of medicine which I have not mentioned, and which is not much practiced in this country, has recently come into being. I refer to the colloidal system. Although even here the methods of preparation have not been pushed quite far enough, the results of some of its experiments would seem to indicate that this particular branch of research work is being conducted on the right lines, and is paving the way to a more efficient system of medicine.

The Rockefeller Institute, in the course of its research work, has demonstrated that iron taken in this form is much more easily absorbed by the body than in its cruder state, whilst copper administered

as a colloidal preparation is a powerful agent in
the reduction of neuralgic and nervous conditions.
In their laboratory experiments too, it has been
found that flowers rescued from the rubbish heap and
placed in a bowl of colloidal copper regain their
freshness.

A further proof of the efficacy of the system was
provided during a bad outbreak of goitre in one of
the American states. The epidemic was almost
entirely eliminated by the addition of a colloidal
preparation of iodine to the supply of drinking-
water in those districts where the goitre was most
prevalent.

For a medicament to be brought to its highest grade
of action, the preparation is of inestimable
importance, but so long as the physician is content
with the preparation of the chemist, I fail to see
how any vital improvement in the quality and
efficacy of our healing mediums can be expected. The
physician is no chemist, the chemist has no clinical
experience, and so the medicinal art must fail
repeatedly not because its students themselves are
incompetent, but because the system under which they
work is so inadequate. We contribute enormous sums
of money to the maintenance of our hospitals and at
the same time drive into them the victims of our
foolish system of drugging and feeding. I repeat, it
is not the body of men that I condemn, but merely
our absurd system of contradictions. Paracelsus has
said:

'If, then, it be of such vast importance that
Alchemy shall be thoroughly understood in Medicine,
the reason of this importance arises from the great

latent virtue which resides in natural things, which also can lie open to none, save insofar as they are revealed by Alchemy. Otherwise it is just as if one should see a tree in winter and not recognize it, or be ignorant what was in it until summer puts forth, one after another, now branches, now flowers, now fruits, and whatever appertains to it. So in these matters there is a latent virtue which is occult to men in general. And unless a man learns and makes proof of these things, which can only be done by an alchemist, just as by the summer, it is not possible that he can investigate the subject in any other way.'

Again he says:

'Who will deny that even in the very best things a poison may be hid? All must acknowledge this. And if this be true, I would now ask you whether it is not right that the poison should be separated from what is good and useful, that the good should be taken and the evil left. Such should certainly be the case. If so, tell me how it is separated in your surgeries. With you all these elements remain mixed. See your own simplicity, then, if you are forced to confess that a poison lies hid, and are asked how it is to be got rid of. Then you bring forth I know not how many correctives, which shall drive out and take away the poison. Does not the poison remain afterwards as before? And yet you boast that you have so corrected it that the poison no longer harms. Whither has it gone? Exceed the proper dose, and you will soon see where the poison is.

'The elimination of a poison can only be done by separation; if this is not brought about you cannot

be sure of your work. If a sure foundation be
necessary for the extraction of the poison, this is
afforded by alchemy. But when the bodies are
contrary, it is absolutely necessary that one of
them should be taken away and removed, so that in
this way all contrariety should be separated from
the good. It is necessary that everything which is
to benefit man shall have passed by fire to a second
birth. Should not this then be deemed the right
fundamental principle by every physician?'

I put forward these ideas because I believe that in
the medicine of metals there is a perfect curative
system; that in the seven metals, gold, silver,
iron, copper, tin, mercury, and lead can be found
elements to cure all discords in the human body, and
that when this system is properly understood and
practised, the multitude of remedies may be
discarded. Be it understood that this is not my
system, but one which is as old as man himself.
Truly it has been said that there is nothing new
under the sun, for knowledge is revealed and is
submerged again, even as a nation rises and falls.
Here is a system, tested throughout the ages, but
lost again and again by ignorance or prejudice, in
the same way that great nations have risen and
fallen and been lost to history beneath the desert
sands and in the ocean depths.

To what end do we study history if not to learn from
it? To profit by the example of those who have gone
before, to learn from their mistakes, if needs be?
Our civilization of today might be a far greater
civilization if it would but borrow from the past,
for knowledge there has always been, and wise men
there have always been, who despite the persecution

and opposition of their fellow men, have yet
laboured to preserve these secrets for posterity.

CHAPTER II: PRACTICAL

In writing this section on the practical work I wish my readers to realize that I am writing purely from the alchemist's, not the chemist's viewpoint. I fully realized when commencing this work that my only hope of success was to put on one side for the time being any knowledge of chemistry that I might possess and to study alchemystical writings in a sincere attempt to understand the alchemist's language and reasoning, and then, by following out his instructions faithfully step by step, to prove the practicability of this science.

The chemist who may read this book must therefore appreciate this point, and understand that at the moment I am not trying to reconcile my findings with the precepts of orthodox chemistry, but merely placing on record my work as an alchemist.

The practice of alchemy in the laboratory has been a far from easy task, as those who have at any time studied literature on the subject will fully appreciate. It is only by continuous experiment and constant comparison with alchemystic writings that the present results have eventually been attained, and looking back on the years of persistence in the face of the countless difficulties and failures which ever confront the would-be alchemist, one can well question the wisdom of pursuing such a course. At last, however, it does seem that these labours may not have been entirely in vain, for from these experiments has gradually emerged the vision of the benefit this art could be to man who, in his present state of imperfection, with its accompanying suffering of mind and body, would seem to require some assistance on his way through life.

As I have said, I believe that in this art lies man's salvation from sickness and disease, and the secret of his ultimate perfection, but needless to say in order to utilize to the full the physical benefits of alchemistic research, man must undertake the transmutation of certain baser elements in his emotional and mental make-up. With this process of psychological transmutation I do not propose to deal for the moment, but I am convinced that in this present age of chaos, when new ideas, new values, and, as I believe, new understanding are coming into being, it may be possible that some of these more unorthodox conceptions will meet with less opposition and more sympathy than previously. Since the complete destruction of all those conditions which in the nineteenth century seemed so permanent and immovable, man has been far less inclined to reject out of hand any new idea which may be put before him. For this reason I write down my findings of an age-old truth in the belief that it is a task destiny has set me, and whether my words be accepted or no lies not with me but with those to whom they are addressed.

Come with me, therefore, to my little laboratory with its array of alembics, crucibles, and sandbaths, and hear something of the struggles of the would-be alchemist and of the mysteries he seeks to unravel.

After a careful study of Basil Valentine's 'Triumphal Chariot of Antimony,' I decided to make my first experiments with antimony. I soon found, however, that on arriving at a crucial point, the key had almost invariably been deliberately withheld, and a dissertation on theology inserted in its place. Gradually, however, I came to realize that the theological discourse was not without object, but actually the means of veiling a valuable clue of some kind. After much labour, a fragrant golden liquid was finally obtained from the antimony, although this was merely a beginning. The alkahest of the alchemist, the First Matter, still remained a mystery.

Then followed processes with iron and copper. After purification of the salts or vitriol of these metals, of calcination, and the obtaining of a salt from the calcined metal by a special process, followed by careful distillation and re-distillation in rectified spirits of wine, the oil of these metals was obtained, a few drops of which used singly, or in conjunction, proved very efficacious in cases of anemia and debility which the ordinary iron medicine failed to touch.

The conjunction of iron and copper proved to be an elixir of a very stimulating and regenerating character, the action being such as to clear the body from toxins, and I well remember on taking a few drops one evening that the prospect of a spell of fairly strenuous mental work, even after a really laborious day, seemed to hold no terrors for me!

But still the alkahest remained an enigma, and so further experiments were made with silver and mercury. For those with silver, fine silver was reduced with nitric acid to the salts of the metal, carefully washed in distilled water, sublimated by special process, finally yielding up a white oil which had a very soothing effect on highly nervous cases.

In the case of mercury, the metal on being reduced to its oil, produced a clear crystalline liquid with great curative properties, but unlike common mercury, no poisonous qualities.

After this I decided to work upon fine gold--gold, that is, without any alloy. This was dissolved in Aqua Regia and reduced to the salts of gold; these were washed in distilled water, which in its turn was evaporated in order to remove its very caustic properties. It was at this point that a very real difficulty arose, for when these salts of gold lose their acidity, they slowly but surely tend to return to their metallic form again. Nevertheless, an elixir was finally produced from them by distillation, although even then a residue of fine metallic gold remained behind in the retort. Having got so far I realized that without the alkahest of the philosophers the real oil of gold could not be obtained, and so again I went back and forth in the alchemists' writings to obtain the clue. The experiments which I had already made considerably lightened my task, and one day while sitting quietly in deep concentration the solution to the problem was revealed to me in a flash, and at the same time many of the enigmatical utterances of the alchemists were made clear.

Here, then, I entered upon a new course of experiment, with a metal for experimental purposes with which I had had no previous experience. This metal, after being reduced to its salts and undergoing special preparation and distillation, delivered up the Mercury of the Philosophers, the Aqua Benedicta, the Aqua Celestis, the Water of Paradise. The first intimation I had of this triumph was a violent hissing, jets of vapour pouring from the retort and into the receiver like sharp bursts from a machine-gun, and then a violent explosion, whilst a very potent and subtle odour filled the laboratory and its surroundings. A friend has described this odour as resembling the dewy earth on a June morning, with the hint of growing flowers in the air, the breath of the wind over heather and hill, and the sweet smell of the rain on the parched earth.

Nicholas Flamel, after searching and experimenting from the age of twenty, wrote when he was eighty years old:

'Finally I found that which I desired, which I also soon knew by the strong scent and odour thereof.'

Does this not coincide, this voice from the fourteenth century, with my own description of the peculiar subtle odour? Cremer, also writing in the early fourteenth century, says:

'When this happy event takes place, the whole house will be filled with a most wonderful sweet fragrance, and then will be the day of the nativity of this most blessed preparation.'

Having arrived at this point my next difficulty was to find a way of storing this subtle gas without danger to property. This I accomplished by coils of glass piping in water joined up with my receiver, together with a perfect government of heat, the result being that the gas gradually condensed into a clear golden-coloured water, very inflammable and very volatile. This water had then to be separated by distillation, the outcome being the white mercurial water described by the Comte St. Germain as his athoeter or primary water of all the metals. I will again quote from Manly Hall's introduction to 'The Most Holy Trinosophia,' the passage in which Casanova describes the athoeter:

'Then he showed me his rnagistrum which he called *Athoeter*. lit was a white liquid contained in a well stopped phial. He told me that this liquid was the universal spirit of Nature and that if the wax of the stopper was pricked ever so slightly, the whole of the contents would disappear. I begged him to make the experiment. He thereupon gave me the phial and the pin and I myself pricked the wax, when, lo, the phial was empty.'

This passage aptly describes this water which is so volatile that it rapidly evaporates if left unstoppered, boils at a very low temperature, and does not so much as wet the fingers. This mercurial water, this athoeter of St. Germain, is absolutely necessary to obtain the oil of gold, which is obtained by its addition to the salts of gold after those salts have been washed with distilled water several times to remove the strong acidity of the Aqua Regia used to reduce the metal to that state. When the Mercurial Water is added to these salts of gold, there is a slight hissing, an increase in heat, and the gold becomes a deep red liquid, from which is obtained, by means of distillation, the oil of gold, a deep amber liquid of an oily consistency.

This oil, which is the potable gold of the alchemist, never returns to the metallic form of gold. I can understand now, I think, how it is that some of the patients to whom Salts of Gold injections have been administered have succumbed to gold poisoning. So long as the salts are in an acid solution, they remain soluble, but directly the dissolving medium loses its acidity and becomes neutral or alkaline, the salts tend to form again into metallic gold. This is probably what happens in the case of the injection of gold salts into the alkaline intercellular fluids, which in some cases leads to fatal results.

Do not imagine that chemists know all about metals! They do not, as the following quotation from the report of Professor Charles Gibson's presidential address on 'Recent Investigations in the Chemistry of Gold' would seem to show:

'The address was of a highly technical nature. One of the chief points brought forward was that current text-book views of the constitution of salts of gold are incorrect. These are never of the same nature as normal metallic salts with simple formulae such as $AuCl$ or $AuBr_3$, but always of a complex constitution. . ."

From the golden water I have described can be obtained this white water, and a deep red tincture which deepens in colour the longer it is kept; these two are the mercury and the sulphur described by the alchemists, Sol the Father and Lune the Mother, the Male and the Female Principles, the White and Red Mercuries, which two conjoined again form a deep amber liquid. This is the *Philosophic Gold,* which is *not* made from metallic gold, but from another metal, and is a *far more Potent* Elixir than the oil of gold. This deep amber liquid literally shines and reflects and intensifies rays of light to an extraordinary degree. It has been described by many alchemists, which fact again corroborates my work in the laboratory. Indeed, every step which I have taken in the laboratory I have found in the work of the various followers of the Spagyric Art.

And now to the final goal, the Philosophers' Stone. Having found my two principles, the Mercury and the Sulphur, my next step was to purify the dead body of the metal, that is, the black dregs of the metal left after the extraction of the golden water. This was calcined to a redness and carefully separated and treated until it became a white salt. The three principles were then conjoined in certain exact quantities in a hermetically sealed flask in a fixed heat neither too hot nor too cold, care as to the exact degree of heat being essential, as any carelessness in its regulation would completely spoil the mixture.

On conjunction the mixture takes on the appearance of a leaden mud, which rises slowly like dough until it throws up a crystalline formation rather like a coral plant in growth. The 'flowers' of this plant are composed of petals of crystal which are continually changing in colour. As the heat is raised, this formation melts into an amber-coloured liquid which gradually becomes thicker and thicker until it sinks into a black earth on the bottom of the glass. At this point (the Sign of the Crow in alchemical literature) more of the ferment or mercury is added. In this process, which is one of continual sublimation, a long-necked, hermetically sealed flask is used, and one can watch the vapour rising up the neck of the flask and condensing down the sides. This process continues until the state of 'dry blackness' is attained. When more of the mercury is added, the black powder is dissolved, and from this conjunction it seems that a new substance is born, or, as the early alchemists would have expressed it, a Son is born. As the black colour abates, colour after colour comes and goes until the mixture becomes white and shining; the White Elixir. The heat is gradually raised yet more, and from white the colour changes to citrine and finally to red--the Elixir Vitae, the Philosophers' Stone, the medicine of men and metals. From their writings, it appears that many alchemists found it unnecessary to take the Elixir to this very last stage, the citrine coloured solution being adequate for their purpose. It is of interest to note that an entirely different manifestation comes into being after the separation of the three elements and their re-conjunction under the sealed vase of Hermes. By the deliberate separation and unification of the Mercury, Sulphur, and Salt, the three elements appear as a more perfect manifestation than in the first place.

CONCLUSION

Man's work is not merely to exist on this earth, to scratch ignorantly at its surface, to mutilate Nature in every possible way, to fight and rob his neighbour, but to develop the powers surrounding him, to manipulate those forces that he may truly and deservedly claim his right to inherit the earth. A garden which has been neglected for years and is overgrown with weeds, when taken over by an intelligent human being who will work hand in hand with nature, may once again become a thing of beauty and joy. Thus the earth, which is man's garden, must be sown and cultivated by him, perfected by his art.

Life is not a haphazard game of chance, but an unfoldment and development of its own powers manifesting in perfect Law. Let us, then, try to understand this Life which is Eternal Law, pervaded by an Intelligence with Order and Wisdom, and having understood, let us work for the more perfect unfoldment of our earth and the forces which lie beneath its surface; for this Law applies to agriculture, to science, to the production of food, to the use of minerals and metals, to the building of cities, to the use of electricity and all natural forces. When man finally learns to use these forces, he will be able to press forward and onward to the final goal, which is the perfection of the earth and of his own species.

Alchemy brings us the vision of the heights to which man may attain; it teaches us that he is Triune, that is, Spiritual, Mental, and Physical; that his

future is far greater than at present can be
envisaged; that, Life is Law and Wisdom.

Those of you who have followed me thus far may be
interested in the following extracts of Hermetic
literature, both of which, apart from their
intrinsic beauty, provide perfect examples of the
highly mystical and intentionally enigmatic
phraseology of alchemical writing.

The authorship of the first, the Tractatus Aureus or
Golden Treatise of Hermes, is unknown, despite the
name it bears. It is, however, thought to be one of
the most ancient and complete pieces of alchemical
writing left to us, and has been held in high esteem
by alchemists of all ages as a complete exposition
of their art.

The second, the Book of the Revelation of Hermes,
interpreted by Theophrastus Paracelsus, concerning
the Supreme Secret of the World, was first published
under the auspices of Benedictus Figulus in his
'Golden and Blessed Casket of Nature's Marvels,' in
1608 (a translation of which work was edited and
introduced by Mr. Arthur Edward Waite in the latter
part of the last century). Many of the truths
enunciated therein are to be found in other works by
writers of earlier and later times, but much of the
phraseology is unique to Paracelsus himself.

'AUREUS,' OR THE GOLDEN TRACTATE

SECTION I

EVEN thus saith Hermes:

"Through long years I have not ceased to experiment, neither have I spared any labour of mind, and this science and art I have obtained by the sole inspiration of the Living God, who judged fit to open them to me His servant, who has given to rational creatures the power of thinking and judging aright, forsaking none or giving to any occasion to despair. For myself, I had never discovered this matter to anyone had it not been from fear of the judgment and the perdition of my soul, if I concealed it. It is a debt which I am desirous to discharge to the faithful as the Father of the faithful did liberally bestow it upon me.

"Understand ye then, O Sons of Wisdom, that the knowledge of the four elements of the ancient philosophers was not corporally or imprudently sought after, which are through patience to be discovered according to their causes and their occult operation. But, their operation is occult, since nothing is done except the matter be decompounded and because it is not perfected unless the colours be thoroughly passed and accomplished. Know then, that the division that was made upon the water, by the ancient philosophers, separates it into four substances, one into two, and three into one, the third part of which is colour, as it were--

a coagulated moisture; but the second and third
waters are the Weights of the Wise.

"Take of the humidity, or moisture, an ounce and a
half, and of the Southern Redness, which is the soul
of gold, a fourth part, that is to say, half an
ounce; of the citrine Seyre, in like manner, half an
ounce; of the Auripigment, half an ounce, which are
eight; that is three ounces. And know ye that the
vine of the wise is drawn forth in three, but the
wine thereof is not perfected, until at length
thirty be accomplished.

"Understand the operation, therefore. Decoction
lessens the matter, but the tincture augments it,
because Luna in fifteen days is diminished, and in
the third she is augmented. This is the beginning
and the end. Behold, I have declared that which was
hidden, since the work is both with thee and about
thee; that which was within is taken out and fixed,
and thou canst have it either in earth or sea.

"Keep, therefore, the Argent vive, which is prepared
in the innermost chamber in which it is coagulated;
for that is the Mercury which is celebrated from the
residual earth.

"He, therefore, who now hears my words, let him
search into them, which are to justify no evil-doer,
but to benefit the good; therefore I have discovered
all things that were before hidden concerning this
knowledge, and disclosed the greatest of all
secrets, even the Intellectual Science.

"Know ye, therefore, Children of Wisdom, who inquire concerning the report thereof, that the vulture standing upon the mountain crieth out with a loud voice: 'I am the White of the Black, and the Red of the White, and the Citrine of the Red, and behold I speak the very Truth.'

"And know that the chief principle of the art is the Crow, which is the blackness of the night and the clearness of the day, and flies without wings. From the bitterness existing in the throat the tincture is taken, the red goes forth from his body, and from his back is taken a thin water.

"Understand, therefore, and accept this gift of God which is hidden from the thoughtless world. In the caverns of the metals there is hidden the stone that is venerable, splendid in colour, a mind sublime and an open sea. Behold, I have declared it unto thee; give thanks to God who teacheth thee this knowledge, for He in return recompenses the grateful.

"Put the matter into a moist fire, therefore, and cause it to boil, in order that its heat may be augmented, which destroys the siccity of the incombustible nature, until the radix shall appear; then extract the redness and the light parts, till only about a third remains.

"Sons of Science! For this reason are philosophers said to be envious, not that they grudged truth to religious or just men, or to the wise, but to fools, ignorant and vicious, who are without Self-Control and benevolence, lest they should be made powerful, and able to perpetrate sinful things. For of such the philosophers are made accountable to God, and evil men are not admitted worthy of this wisdom.

"Know that this matter I call the stone, but it is also named the feminine of magnesia, or the hen, or the white spittle, or the volatile milk, the incombustible oil, in order that it may be hidden from the inept and ignorant, who are deficient in goodness and self-control; which I have nevertheless signified to the wise by one only epithet, viz., the Philosophers' Stone.

"Include, therefore, and conserve in this sea, the fire, and the heavenly bird, to the latest moment of his exit. But I deprecate ye all, Sons of Philosophy, on whom the great gift of this knowledge being bestowed, if any should undervalue or divulge the power thereof to the ignorant, or such as are unfit for the knowledge of this secret. Behold, I have received nothing from any to whom I have not returned that which had been given me, nor have I failed to honour him; even in this I have reposed the highest confidence.

"This, O Son, is the concealed Stone of many colours, which is born and brought forth in one colour; I know this and conceal it. By this, the Almighty favouring, the greatest diseases are escaped, and every sorrow, distress and evil and hurtful thing is made to depart; for it leads from darkness into light, from this desert wilderness to a secure habitation, and from poverty and straits to a free and ample fortune."

SECTION II

"My son, before all things I admonish thee to fear God, in whom is the strength of thy undertaking, and the bond of whatsoever thou meditatest to unloose; whatsoever thou hearest, consider it rationally. For I hold thee not to be a fool. Lay hold, therefore, of my instructions and meditate upon them, and so let thy heart be fitted also to conceive, as if thou was thyself the author of that which I now teach. If thou appliest cold to any nature that is hot, it will not hurt it; in like manner, he who is rational shuts himself within from the threshold of ignorance, lest supinely he should be deceived.

"Take the flying bird and drown it flying, and divide and separate it from its pollutions, which yet hold it in death; draw it forth and repel it from itself, that it may live and answer thee, not by flying away into the regions above but by truly forbearing to fly. For if thou shalt deliver it out of its prison, after this thou shalt govern it according to Reason, and according to the days that I shall teach thee: then will it become a companion unto thee, and by it thou wilt become to be an honoured lord.

"Extract from the ray its shadow, and from the light its obscurity, by which the clouds hang over it and keep away the light: by means of its construction, also, and fiery redness, it is burned.

"Take, my Son, this redness, corrupted with water, which is as a live coal holding fire, which if thou shalt withdraw so often until the redness is made pure, then it will associate with thee, by whom it was cherished, and in whom it rests.

"Return, then, O my Son, the coal being extinct in life, upon the water for thirty days, as I shall note to thee, and henceforth thou art a crowned king, resting over the fountain, and drawing from thence Auripigment dry without moisture. And now I have made the heart of the hearers, hoping in thee, to rejoice, even in their eyes, beholding thee in anticipation of that which thou possessest.

"Observe, then, that the water was first in the air, then in the earth; restore thou it also to the superiors by its proper windings and not foolishly altering it; then to the former spirit, gathered in its redness, let it be carefully conjoined.

"Know, my Son, that the fatness of our earth is sulphur, the auripigment sirety, and colcothar which are also sulphur, of which auripigments sulphur, and such like, some are more vile than others, in which there is a diversity, of which kind also is the fat of gluey matters, such as are hair, nails, hoofs, and sulphur itself, and of the brain, which too is auripigment, of the like kind also are the lions' and cats' claws, which is sirety the fat of white bodies, and the fat of the two oriental quicksilvers, which sulphurs are hunted and retained by the bodies.

"I say, moreover, that this sulphur doth tinge and fix, and is held by the conjunction of the tinctures; oils also tinge, but fly away, which in the body are contained, which is a conjunction of fugitives only with sulphurs and albuminous bodies, which hold also and detain the fugitive ens.

"The disposition sought after by the philosophers, O Son, is but one in our egg, but this in the hen's egg is much less to be found. But lest so much of the Divine Wisdom as is a hen's egg should not be distinguished, our composition is, as that is, from the four elements adapted and composed. Know, therefore, that in the hen's egg is the greatest help with respect to the proximity and relationship of the matter in nature for in it there is a spirituality and conjunction of elements, and an earth which is golden in its tincture."

But the Son, inquiring of Hermes, saith:
"The sulphurs which are fit for our work, whether they are celestial or terrestrial?"

To whom the Father replies:

"Certain of them are heavenly and some are of the earth."

Then the Son saith:

"Father, I imagine the heart in the superiors to be heaven, and in the inferiors, earth."

But saith Hermes:

"It is not so; the masculine is truly the heaven of the feminine, and the feminine is the earth of the masculine."

The Son then asks:

"Father, which of these is more worthy than the other, whether is it the heaven or the earth?"

Hermes replies:

"Both need the help one of the other, for the precepts demand a medium."

But saith the Son:

"If thou shalt say that a wise man governs all mankind?"

"But ordinary men," replies Hermes, "are better for them, because every nature delights in society of its own kind, and so we find it to be in the life of Wisdom where equals are conjoined."

"But what," rejoins the Son, "is the mean betwixt them?"

To whom Hermes replies:

"In everything in nature there are three from two; the beginning, the middle, and the end. First the needful water, then the oily tincture, and lastly, the faeces, or earth, which remains below.

"But the Dragon inhabits in all these, and his houses are the darkness and blackness that is in them, and by them he ascends into the air, from his rising, which is their heaven. But whilst the fume remains in them, they are not immortal. Take away, therefore, the vapour from the water, and the blackness from the oily tincture, and death from the faeces, and by dissolution thou shalt possess a triumphant reward, even that in and by which the possessors live.

"Know then, my Son, that the temperate unguent, which is fire, is the medium between the faeces and the water, and is the Perscrutinator of the water. For the unguents are called sulphurs, because between fire and oil and this sulphur there is such a close proximity, that even as fire burns so does the sulphur also.

"All the sciences of the world, O Son, are comprehended in this my hidden Wisdom, and this, and the learning of the Art, consists in these wonderful hidden elements which it doth discover and complete. It behoves him, therefore, who would be introduced to this hidden Wisdom, to free himself from the hidden usurpations of vice, and to be just and good and of a sound reason, ready at hand to help mankind, of a serene countenance, diligent to save, and be himself a patient guardian of the arcane secrets of philosophy.

"And this know, that except thou understandest how to mortify and induce generation, to vivify the Spirit and introduce Light, until they fight each other and grow white and freed from their defilements, rising as it were from blackness and darkness, thou knowest nothing nor canst perform anything. But if thou knowest this, thou wilt be of a great dignity so that even kings themselves shall reverence thee. These secrets, Son, it behoves thee to conceal from the vulgar and profane world.

"Understand, also, that our Stone is from many things and of various colours, and composed from four elements which we ought to divide and dissever in pieces, and segregate, in the veins, and partly mortifying the same by its proper nature, which is also in it, to preserve the water and fire dwelling therein, which is from the four elements and their waters, which contain its water; this, however, is not water in its true form, but fire, containing in a pure vessel the ascending waters, lest the spirits should fly away from the bodies; for by this means they are made tingeing and fixed.

"O, blessed watery form, that dissolvest the elements! Now it behoves us, with this watery soul, to possess ourselves of a sulphurous form, and to mingle the same with our Acetum. For when, by the power of water, the composition is dissolved, it is the key of the restoration; then darkness and death will fly away from them and Wisdom proceeds onwards to the fulfilment of her Law."

SECTION III

"Know, my Son, that the <u>philosophers</u> bind up their matter with a strong chain that it may contend with the Fire; because the spirits in the washed bodies desire to dwell therein and to rejoice. In these habitations they vivify themselves and inhabit there, and the bodies hold them, nor can they be hereafter separated any more.

"The dead elements are revived, the composed bodies tinge and are altered, and by a wonderful process they are made permanent, as saith the philosopher. "O, permanent watery Form, creatrix of the royal elements! who, having with thy brethren and a just government obtained the tincture, findest rest. Our precious stone is cast forth upon the dung-hill, and that which is most worthy is made vilest of the vile. Therefore, it behoves us to mortify two Argent vives together, both to venerate and be venerated, viz., the Argent vive of Auripigment, and the oriental Argent vive of Magnesia.

"O, Nature, the most potent creatrix of Nature, which containest and separatist natures in a middle principle. The Stone comes with light, and with light it is generated, and then it generates and brings forth the black clouds of darkness, which is the mother of all things.

"But when we marry the crowned King to our red daughter, and in a gentle fire, not hurtful she doth Conceive an excellent and supernatural son, which permanent life she doth also feed with a subtle heat, so that he lives at length in our fire.

"But when thou shalt send forth thy fire upon the foliated sulphur, the boundary of hearts doth enter in above, it is washed in the same, and the purified matter thereof is extracted.

"Then he is transformed, and his tincture by help of the fire remains red, as it were flesh. But our Son, the king begotten, takes his tincture from the fire, and death even, and darkness, and the waters flee away.

"The Dragon shuns the sunbeams which dart through the crevices and our dead son lives; the king comes forth from the fire and rejoins with his spouse, the occult treasures are laid open, and the virgin's milk is whitened. The Son, already vivified, is become a warrior in the fire, and of tincture super-excellent. For this Son is himself the treasury, even himself bearing the Philosophic Matter.

"Approach, ye Sons of wisdom, and rejoice; let us now rejoice together, for the reign of death is finished, and the Son doth rule. And he is invested with the red garment, and the scarlet colour is put on."

SECTION IV

"Understand, then, O Son of Wisdom, what the Stone declares: 'Protect me and I will protect Thee; increase my strength that I may help thee! My Sol and my beams are most inward and secretly in me, my own Luna, also, is my light, exceeding every other light, and my good things are better than all other good things, I give freely, and reward the intelligent with joy and gladness, glory, riches, and delights, and them that seek after me I make to know and understand, and to possess divine things.'

"Behold, that which the philosophers have concealed is written with seven letters; for Alpha and Yda follow two, and Sol in like manner follows the book. Nevertheless, if thou art willing that he should have Dominion, observe the Art, and join the son to the daughter of the water, which is Jupiter and a hidden secret.

"Auditor, understand. Let us use our Reason. Consider all with the most accurate investigation, which in the contemplative part I have demonstrated to thee, the whole matter I know to be the one only thing. But who is he that understands the true investigation and inquires rationally into this matter? It is not from man, nor from anything like him or akin to him; nor from the ox or bullock, and if any creature conjoins with one of another species, that which is brought forth is neutral from either."

"Thus saith Venus: 'I beget light, nor is the darkness of my nature, and if my metal be not dried all bodies desire me, for I liquefy them and wipe away their rust, even I extract their substance. Nothing, therefore is better or more venerable than I, my brother also being conjoined.'

"But the King, the Ruler, to his brethren, testifying of him, saith: 'I am crowned, and I am adorned with a royal diadem. I am clothed with the royal garment, and I bring joy and gladness of heart, for being chained, I caused my substance to lay hold of, and to rest within the arms and breast of my mother, and to fasten upon her substance, making that which was invisible to become visible, and the occult matter to appear. And everything which the philosophers have hidden is generated by us. Hear, then, these words, and understand them. Keep them, and meditate thereon, and seek for nothing more. Man in the beginning is generated of nature, whose inward substance is fleshy, and not from anything else. Meditate on these plain things, and reject what is superfluous.'

"Thus saith the philosopher: 'Botri is made from the citrine, which is extracted out of the Red Root, and from nothing else; and if it be citrine and nothing else Wisdom was with thee. It was not gotten by thy care, nor if it be freed from redness, by thy study. Behold, I have circumscribed nothing. If thou hast understanding, there be but few things unopened.

"Ye Sons of Wisdom! Turn then the Breym Body with an exceeding great fire, and it will yield gratefully what you desire. And see that you make that which is volatile, so that it cannot fly, and by means of that which flies not. And that which yet rests upon the fire, as it were itself a fiery flame, and that which in the heat of a boiling fire is corrupted, is cambar.

"And know ye that the Art of this permanent water is our brass and the colouring of its tincture and blackness is then changed into the true red.

"I declare that, by the help of God, I have spoken nothing but the truth. That which is destroyed is renovated, and hence the corruption is made manifest in the matter to be renewed, and hence the melioration will appear, and on either side it is a signal of Art."

SECTION V

"My Son, that which is born of the Crow is the beginning of this Art. Behold, now I have obscured the matter treated of, by circumlocution, depriving thee of the light. Yet this dissolved, this joined, this nearest and farthest off, I have named to thee. Roast those things, therefore, and boil them in that which comes from the horse's belly for seven, fourteen or twenty-one days. Then will the Dragon eat his own wings and destroy himself. This being done, let it be put into a fiery furnace, which lute diligently, and observe that none of the spirit may escape.

"And know that the periods of the earth are in the water, which let it be as long as until thou puttest the same upon it. This matter being thus melted and burned, take the brain thereof and triturate it in most sharp vinegar, till it becomes obscured. This done, it lives in the putrefaction, let the dark clouds which were in it before it was killed be converted into its own body. Let this process be repeated, as I have described, let it again die, as I before said, and then it lives.

"In the life and death thereof we work with the spirits, for as it dies by the taking away of the spirit, so it lives in the return and is revived and rejoices therein. Being arrived then at this knowledge, that which thou hast been searching for is made apparent in the Affirmation. I have even related to thee the joyful signs, even that which doth fix the body. But these things, and how they attained to the knowledge of this secret, are given by our ancestors in figures and types. Behold, they are dead. I have opened the riddle, and the book of knowledge is revealed. The hidden things I have uncovered, and have brought together the scattered truths within their boundary, and have conjoined many various forms; even I have associated the spirit. Take it as the gift of God."

SECTION VI

"It behoves thee to give thanks to God, Who has bestowed liberally of his bounty to the Wise, Who delivers us from misery and poverty. I am tempted and proven with the fullness of His substance and His probable wonders, and humbly pray God that whilst we live we may come to Him.

"Remove thence, O Sons of Science, the unguents which we extract from fats, hair, verdigrease, tragacanth and bones, which are written in the books of our fathers. But concerning the ointments which contain the tincture, coagulate the fugitive, and adorn the sulphurs, it behoves us to explain their disposition more at large, and to unveil the Form, which is buried and hidden from other unguents, which is seen in disposition, but dwells in his own body, as fire in trees and stones, which by the most subtle art and ingenuity it behoves to extract without burning.

"And know that the heaven is to be joined mediately with the earth, but the Form is in a middle nature between the heaven and the earth, which is our water. But the water holds of all the first place which goes forth from this stone. But the second is gold, and the third is gold, only in a mean which is more noble than the water and the faeces.

"But in these are the smoke, the blackness and the death. It behoves us, therefore, to dry away the vapour from the water, to expel the blackness from the unguent, and death from the faeces and this by dissolution. By which means we attain to the highest philosophy and secret of all hidden things."

SECTION VII

"Know ye then, O Sons of Science, there are seven bodies, of which gold is the first, the most perfect, the king of them, and their head, which neither the earth can corrupt nor fire devastate, nor the water change for its complexion is equalized, and its nature regulated with respect to heat, cold and moisture; nor is there anything in it which is superfluous, therefore the philosophers do buoy up and magnify themselves in it, saying that this gold, in relation to other bodies is, as the sun amongst the stars, more splendid in Light; and as, by the power of God, every vegetable and all the fruits of the earth are perfected, so gold by the same power sustaineth all.

"For as dough without a ferment cannot be fermented so when thou sublimest the body and purifiest it, separating the uncleanness from it, thou wilt then conjoin and mix them together, and put in the ferment confecting the earth and water. Then will the Ixir ferment even as dough doth ferment. Think of this, and see how the ferment in this case doth change the former natures to another thing. Observe also, that there is no ferment otherwise than from the dough itself.

"Observe, moreover, that the ferment whitens the confection and hinders it from turning, and holds the tincture lest it should fly, and rejoice the bodies, and makes them intimately to join and to enter one into another, and this is the key of the philosophers and the end of their work, and by this science, bodies are meliorated, and the operation of them, God assisting, is consummate.

"But, through negligence and a false opinion of the matter, the operation may be perverted, as a mass of leaven growing corrupt, or milk turned with rennet for cheese, and musk among aromatics.

"The sure colour of the golden matter for the red, and the nature thereof, is not sweetness; therefore we make of them sericum--i.e., Ixir; and of them we make the enamel of which we have already written, and with the king's seal we have tinged the clay, and in that have set the colour of heaven, which augments the sight of them that see.

"The Stone, therefore, is the most precious gold without spots, evenly tempered, which neither fire, nor air, nor water, nor earth is able to corrupt; for it is the Universal Ferment rectifying all things in a medium composition, whose complexion is yellow and a true citrine colour.

"The gold of the wise, boiled and well digested, with a fiery water, makes Ixir, for the gold of the wise is more heavy than lead, which in a temperate composition is a ferment Ixir, and contrariwise, in our intemperate composition, is the confusion of the whole.

"For the work begins from the vegetable, next from the animal, as in a hen's egg, in which is the greatest help, and our earth is gold, all of which we make sericum, which is the ferment Ixir."

THE BOOK OF THE REVELATION OF HERMES

INTERPRETED BY THEOPHRASTUS PARACELSUS
CONCERNING THE SUPREME SECRET OF THE WORLD

Hermes, Plato, Aristotle, and the other philosophers, flourishing at different times, who have introduced the Arts, and more especially have explored the secrets of inferior creation, all these have eagerly sought a means whereby man's body might be preserved from decay and become endued with immortality. To them it was answered that there is nothing which might deliver the mortal body from death; but that there is One Thing which may postpone decay, renew youth, and prolong short human life (as with the Patriarchs). For death was laid as a punishment upon our first parents, Adam and Eve, and will never depart from all their descendants. Therefore, the above philosophers, and many others, have sought this One Thing with great labour, and have found that which preserves the human body from corruption, and prolongs life, conducts itself, with respect to other elements, as it were like the Heavens from which they understood that the Heavens are a substance above the Four Elements. And just as the Heavens, with respect to the other elements are held to be the fifth substance (for they are indestructible, stable, and suffer no foreign admixture), so also this One Thing (compared to the forces of our body) is an indestructible essence, drying up all the superfluities of our bodies, and has been philosophically called by the above-mentioned name. It is neither hot and dry like fire, nor cold and moist like water, nor warm and moist like air, nor dry and cold like earth. But it is a skillful, perfect equation of all the Elements, a right commingling of natural forces, a most particular union of spiritual virtues, an indissoluble uniting of body and soul. It is the purest and noblest substance of an indestructible body, which cannot be destroyed nor harmed by the Elements, and is produced by Art. With this Aristotle prepared an apple prolonging life by its scent, when he, fifteen days before his death, could neither eat nor drink on account of old age. This spiritual Essence, or One Thing, was revealed from above to Adam, and was greatly desired by the Holy Fathers, this also Hermes and Aristotle call the Truth without Lies, the most sure of all things certain, the Secret of all Secrets. It is the Last and the Highest Thing to be sought under the Heavens, a wondrous closing and finish of philosophical work, by which are discovered the dews of Heaven and the fastnesses of Earth. What the mouth of man cannot utter is all found in this Spirit. As Morienus says: 'He who has this has all things, and wants no other aid. For in it are all temporal happiness, bodily health, and earthly fortune. It is the spirit of the fifth substance, a Fount of all Joys (beneath the rays of the moon), the Supporter of Heaven and Earth, the Mover of Sea and Wind, the Outpourer of Rain, upholding the strength of all things, an excellent spirit above Heavenly and other spirits, giving Health, Joy, Peace, Love: driving away Hatred and Sorrow, bringing in Joy, expelling all Evil, quickly healing all Diseases, destroying Poverty and Misery, leading to all good things, preventing all evil words and thoughts, giving man his heart's desire, bringing to the pious earthly honour and long life, but to the wicked who misuse it, Eternal Punishment.'

This is the Spirit of Truth, which the world cannot comprehend without the interposition of the Holy Ghost, or without the instruction of those who know it. The same is of a mysterious nature, wondrous strength, boundless power. The Saints, from the beginning of the world, have desired to behold its face. By Avicenna this Spirit is named the Soul of the World. For as the Soul moves all the limbs of the Body, so also does this Spirit move all bodies. And as the Soul is in all the limbs of the Body, so also is this Spirit in all elementary created things. It is sought by many and found by few. It is beheld from afar and found near; for it exists in everything, in every place, and at all times. It has the powers of all creatures; its action is found in all elements, and the qualities of all things are therein, even in the highest perfection. By virtue of this essence did Adam and the Patriarchs preserve their health and live to an extreme age, some of them also flourishing in great riches.

When the philosophers had discovered it, with great diligence and labour, they straightway concealed it under a strange tongue, and in parables, lest the same should become known to the unworthy, and the pearls be cast before swine. For if everyone knew it, all work and industry would cease; man would desire nothing but this one thing, people would live wickedly, and the world be ruined, seeing that they would provoke God by reason of their avarice and superfluity. For eye hath not seen, nor ear heard, nor hath the heart of man understood what Heaven hath naturally incorporated with this Spirit. Therefore have I briefly enumerated some of the qualities of this Spirit, to the Honour of God, that the pious may reverently praise Him in His gifts (which gift of God shall afterwards come to them), and I will herewith shew what powers and virtues it possesses in each thing, also its outward appearance, that it may be more readily recognized. In its first state, it appears as an impure earthly body, full of imperfections. It then has an earthly nature, healing all sickness and wounds in the bowels of man, producing good and consuming proud flesh, expelling all stench, and healing generally, inwardly and outwardly.

In its second nature, it appears as a watery body, somewhat more beautiful than before, because (although still having its corruptions) its Virtue is greater. It is much nearer the Truth, and more effective in works. In this form it cures cold and hot fevers, and is a specific against poisons, which it drives from heart and lungs, healing the same when injured or wounded, purifying the blood, and, taken three times a day, is of great comfort in all diseases.

But in its third nature it appears as an aerial body of an oily nature, almost freed from all imperfections, in which form it does many wondrous works, producing beauty and strength of body, and (a small quantity being taken in the food) preventing melancholy and heating of the gall, increasing the quantity of blood and seed. It expands the blood vessels, cures withered limbs, restores strength to the sight, in growing persons removes what is superfluous and makes good defects in the limbs.

In its fourth nature it appears in a fiery form (not quite freed from all imperfections, still somewhat watery and not dried enough), wherein it has many virtues making the old young and reviving those at the point of death. For if to such an one there be given, in wine, a barleycorn's weight of this fire, so that it reach the stomach, it goes to his heart, renewing him at once, driving away all previous moisture and poison, and restoring the natural heat of the liver. Given in small doses to old people, it removes the diseases of age, giving the old young hearts and bodies. Hence it is called the Elixir of Life.

In its fifth and last nature, it appears in a glorified and illuminated form, without defects, shining like gold and silver, wherein it possesses all previous powers and virtues in a higher and more wondrous degree. Here its natural works are taken for miracles. When applied to the roots of dead trees they revive, bringing forth leaves and fruit. A lamp, the oil of which is mingled with this spirit, continues to burn forever without diminution. It converts crystals into the most precious stones of all colours, equal to those from the mines, and does many other incredible wonders which may not be revealed to the unworthy.

For it heals all dead and living bodies without other medicine. Here Christ is my witness that I lie not, for all heavenly influences are united and combined therein.

This essence also reveals all treasures in earth and sea, converts all metallic bodies into gold, and there is nothing like unto it under Heaven.

This spirit is the secret, hidden from the beginning yet granted by God to a few holy men for the revealing of these riches to His Glory--dwelling in fiery form in the air, and leading earth with itself to Heaven, while from its body there flow whole rivers of living water. This spirit flies through the midst of the Heavens like a morning mist, leads its burning fire into the water, and has its shining realm in the Heavens.

And although these writings may be regarded as false by the reader, yet to the initiated they are true and possible, when the hidden sense is properly understood. For God is wonderful in His works, and His wisdom is without end.

This spirit in its fiery form is called a Sandaraca, in the aerial a Kybrick, in the watery an Azoth, in the earthly Alcohoph and Aliocosoph. Hence they are deceived by these names, who, seeking without instruction, think to find this Spirit of Life in things foreign to our Art. For although this Spirit which we seek, on account of its qualities, is called by these names, yet the same is not in these bodies and cannot be in them. For a refined spirit cannot appear except in a body suitable to its nature. And, by however many names it be called, let no one imagine there be different spirits, for, say what one will, there is but one spirit working everywhere and in all things.

That is the spirit which, when rising, illumines the Heavens, when setting incorporates the purity of Earth, and when brooding has embraced the Waters. This spirit is named Raphael, the Angel of God, the subtlest and purest, whom the others all obey as their King.

This spiritual substance is neither heavenly nor
hellish, but an airy, pure, and hearty body, midway
between the highest and the lowest, without reason,
but fruitful in works, and the most select and
beautiful of all other heavenly things.

This work of God is far too deep for understanding
for it is the last, greatest, and highest secret of
Nature. It is the Spirit of God, which in the
Beginning filled the Earth and brooded over the
waters, which the world cannot grasp without the
gracious interposition of the Holy Spirit and
instruction from those who know it, which also the
whole world desires for its virtue, and which cannot
be prized enough. For it reaches to the planets,
raises the clouds, drives away mists, gives its
light to all things, turns everything into Sun and
Moon, bestows all health and abundance of treasure,
cleanses the leper, brightens the eyes, banishes
sorrow, heals the sick, reveals all hidden
treasures, and, generally, cures all diseases.
Through this spirit have the philosophers invented
the Seven Liberal Arts, and thereby gained their
riches. Through the same Moses made the golden
vessels in the Ark, and King Solomon did many
beautiful works to the honour of God. Therewith
Moses built the Tabernacle, Noah the Ark, Solomon
the Temple. By this Ezra restored the Law, and
Miriam, Moses' sister, was hospitable; Abraham,
Isaac, and Jacob, and other righteous men, have had
life-long abundance and riches; and all the saints
possessing it have therewith praised God. Therefore
is its acquisition very hard, more than that of gold
and silver. For it is the best of all things,
because, of all things mortal that man can desire in
this world, nothing can compare with it, and in it
alone is truth. Hence it is called the Stone and

Spirit of Truth; in its works is no vanity, its praise cannot be sufficiently expressed. I am unable to speak enough of its virtues, because its good qualities and powers are beyond human thoughts, unutterable by the tongue of man, and in it are found the properties of all things. Yea, there is nothing deeper in Nature.

O unfathomable abyss of God's Wisdom, which thus hath united and comprised in the virtue and power of this one Spirit the qualities of all existing bodies!

O unspeakable honour and boundless joy granted to mortal man! For the destructible things of Nature are restored by virtue of the said Spirit.

O mystery of mysteries, most secret of all secret things, and healing and medicine of all things! Thou last discovery in earthly natures, last best gift to Patriarchs and Sages, greatly desired by the Whole world! Oh, what a wondrous and laudable spirit is purity, in which stand all joy, riches, fruitfulness of life, and art of all arts, a power which to its initiates grants all material joys! O desirable knowledge, lovely above all things beneath the circle of the Moon, by which Nature is strengthened, and heart and limbs are renewed, blooming youth is preserved, old age driven away, weakness destroyed, beauty in its perfection preserved, and abundance ensured in all things pleasing to men! O thou spiritual substance, lovely above all things! O thou wondrous power, strengthening all the world! O thou invincible virtue, highest of all that is, although despised by the ignorant, yet held by the wise in great praise, honour, and glory, that--proceeding

from humours--wakest the dead, expellest diseases,
restorest the voice of the dying!

O thou treasure of treasures, mystery of mysteries,
called by Avicenna 'an unspeakable substance,' the
purest and most perfect soul of the world, than
which there is nothing more costly under Heaven,
unfathomable in nature and power, wonderful in
virtue and works, having no equal among creatures,
possessing the virtues of all bodies under Heaven!
For from it flow the water of life, the oil and
honey of eternal healing, and thus hath it nourished
them with honey and water from the rock. Therefore,
saith Morienus: 'He who hath it, the same also hath
all things.' Blessed art Thou, Lord God of our
Fathers, in that Thou has given the prophets this
knowledge and understanding, that they have hidden
these things (lest they should be discovered by the
blind, and those drowned in worldly godlessness) by
which the wise and pious have praised Thee! For the
discoverers of the mystery of this Thing to the
unworthy are breakers of the seal of Heavenly
Revelation, thereby offending God's Majesty, and
bringing upon themselves many misfortunes and the
punishments of God.

Therefore, I beg all Christians, possessing this
knowledge, to communicate the same to nobody, except
it be to one living in Godliness, of well-proved
virtue, and praising God, Who has given such a
treasure to man. For many seek, but few find it.
Hence the impure and those living in vice are
unworthy of it. Therefore is this Art to be shown to
all God-fearing persons, because it cannot be bought
with a price. I testify before God that I lie not,

although it appear impossible to fools, that no one has hitherto explored Nature so deeply.

The Almighty be praised for having created this Art and for revealing it to God-fearing men. Amen.

And thus is fulfilled this precious and excellent work, called the revealing of the occult spirit, in which lie hidden the secrets and mysteries of the world.

But this spirit is one genius, and Divine, wonderful and lordly power. For it embraces the whole world, and overcomes the Elements and the fifth Substance.

To our Trismegistus Spagyrus,
Jesus Christ,
Be praise and glory immortal.
Amen.

A Word from the Publisher

Thank you for purchasing this small work from The R.A.M.S. Library of Alchemy. During his lifetime, Hans Nintzel was dedicated to the identification, acquisition, study, retyping and, when necessary, translation of what he considered to be the most important known works on Alchemy. Hans was assisted by his sparse network of fellow Alchemists, all members of the Restorers of Alchemical Manuscripts Society (R.A.M.S.). I was an active member of R.A.M.S.

My goal is to publish all of the works originally made available through R.A.M.S. as photocopies. To facilitate this, I have chosen to have the books professionally printed. I also have a few titles that I intend to add to the original R.A.M.S. Library, selected by strict criteria established by Hans.

The works from the original R.A.M.S. Library are republished by R.A.M.S. Publishing Company in the collection, "The R.A.M.S. Library of Alchemy," with permission of the Estate of Hans W. Nintzel.

If you have a work on Alchemy that you believe should be a part of the R.A.M.S. Library, please contact me through R.A.M.S. Publishing Company.

Philip N. Wheeler

www.ingramcontent.com/pod-product-compliance
Lightning Source LLC
Chambersburg PA
CBHW080811180526
45168CB00006B/2410